Four Fish

ALSO BY PAUL GREENBERG

Leaving Katya: A Novel

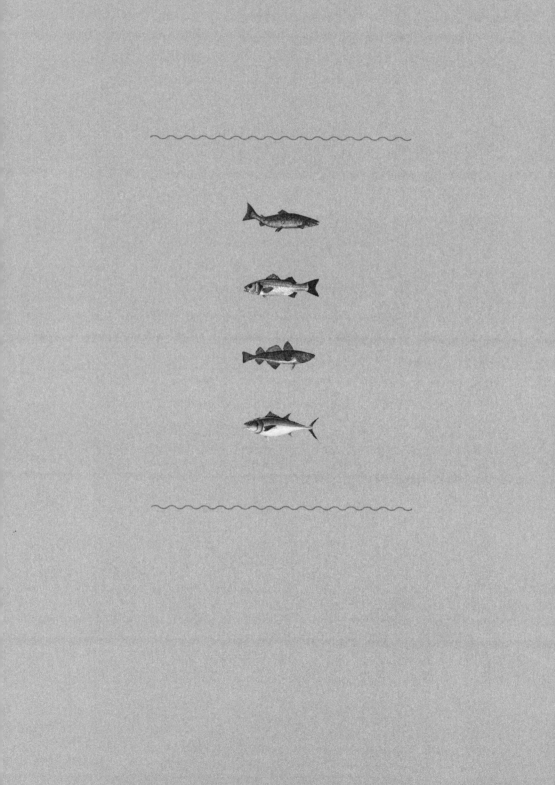

Four Fish

THE FUTURE OF THE
LAST WILD FOOD

·

PAUL GREENBERG

THE PENGUIN PRESS

New York

2010

THE PENGUIN PRESS
Published by the Penguin Group
Penguin Group (USA) Inc., 375 Hudson Street, New York, New York 10014,
U.S.A. • Penguin Group (Canada), 90 Eglinton Avenue East, Suite 700, Toronto,
Ontario, Canada M4P 2Y3 (a division of Pearson Penguin Canada Inc.) •
Penguin Books Ltd, 80 Strand, London WC2R 0RL, England • Penguin Ireland,
25 St. Stephen's Green, Dublin 2, Ireland (a division of Penguin Books Ltd) •
Penguin Books Australia Ltd, 250 Camberwell Road, Camberwell, Victoria 3124,
Australia (a division of Pearson Australia Group Pty Ltd) • Penguin Books
India Pvt Ltd, 11 Community Centre, Panchsheel Park, New Delhi – 110 017,
India • Penguin Group (NZ), 67 Apollo Drive, Rosedale, North Shore 0632,
New Zealand (a division of Pearson New Zealand Ltd) • Penguin Books
(South Africa) (Pty) Ltd, 24 Sturdee Avenue, Rosebank,
Johannesburg 2196, South Africa

Penguin Books Ltd, Registered Offices:
80 Strand, London WC2R 0RL, England

First published in 2010 by The Penguin Press,
a member of Penguin Group (USA) Inc.

Portions of this book appeared in different form in *The New York Times* as "Bass
Market," "Green to the Gills," and "A Catfish by Any Other Name."

Excerpt from *Up in the Old Hotel* by Joseph Mitchell. Copyright © 1992
by Joseph Mitchell. By permission of Pantheon Books, a division of
Random House, Inc.

LIBRARY OF CONGRESS CATALOGING IN PUBLICATION DATA
Greenberg, Paul.
Four fish : the future of the last wild food / by Paul Greenberg.
p.   cm.
Includes index.
ISBN 978-1-59420-256-8
1. Salmon—Cultural control.   2. Sea basses—Cultural control.
3. Codfish—Cultural control.   4. Tuna—Cultural control.   5. Fish culture.
6. Fishery management.   I. Title.
SH167.S17G74 2010
333.95'6—dc22   2010001276

Printed in the United States of America
3   5   7   9   10   8   6   4   2

Designed by Amanda Dewey

For Esther, who knows the depths

CONTENTS

Conclusion   *243*

Epilogue   *257*

Fish is the only grub left that the scientists haven't been able to get their hands on and improve. The flounder you eat today hasn't got any more damned vitamins in it than the flounder your great-great-granddaddy ate, and it tastes the same. Everything else has been improved *and* improved *and* improved to such an extent that it ain't fit to eat.

—a Fulton Fish Market denizen, in *Old Mr. Flood* by Joseph Mitchell, 1944

Four Fish

Introduction

In 1978 all the fish I cared about died. They were the biggest largemouth bass I had ever seen, and they lived in a pond ten minutes' walk from my house on a large estate in the backwoods of Greenwich, Connecticut, perhaps the most famously wealthy town in America. We did not own the house, the estate, the pond, or the largemouth bass, but I still thought of the fish as my fish. I had found them, and the pond was my rightful hunting ground.

My mother had rented the house as she would three other homes in Greenwich, because it gave the illusion of magnificent proprietorship. She tended toward small cottages on large estates—converted stables, liverymen's accommodations that were the unclaimed, declining appendages of older, fading wealth, unsold because of divorces or other family complications, rented out to us for a reasonable fee that would become unreasonable and impel our moving on to other cottages on other collapsing estates.

Fishing was the one constant during these years. Sensing in it a masculine, character-building quality, my mother arranged it so that the cottages we rented always had access to streams and lakes or abutted other properties we could trespass upon that had such resources. She trusted my instincts for spotting fishy water and used me as a kind of divining rod before signing a lease. And for most of my childhood, we were within a short walk of a potentially fruitful cast. Our longest residence was in the aforementioned house near the giant largemouth bass. In the first two years we lived there, I spent all my summer evenings and weekend mornings pursuing them.

In the winter of 1978, though, a fierce blizzard hit southern Connecticut. Temperatures were often below zero and at one point it snowed for thirty-three hours straight. Perhaps it was the cold that killed the fish, or the copper sulfate I helped the caretaker drag through the pond the previous summer to manage the algal blooms, or maybe even the fishermen I'd noticed trespassing on the estate one day, scoping out my grounds. But whatever caused it, after that winter never again did I spot a living fish. Of course I tried. I trolled pretty much every square foot after school the following year, often with a neighbor who had moved in after the era of the great fish. When two months of dragging lures up and down the shoreline produced not even a strike, my neighbor finally stuck a pin in my irrational bubble of hope.

"I don't care what you say about what *was*," I remember him shouting. "There is not a fucking fish in this whole goddamn lake, and I'm *never* fishing here again."

Like any hunter whose grounds have gone bad, I set out looking for new territory. I followed the outflow of the pond down a series of cascades that in turn flattened out into a low, swampy meadow of deep oxbows. Only minuscule shiners, crawfish, and escaped goldfish swam here. Farther and farther I went, until the stream joined a larger river and passage was blocked by a fence that a wealthy landowner had erected. Inspecting a map at the library, I found that this was a significant juncture for my stream (as with "my" pond, I had annexed the stream and referred to it now as "mine"). The point at which it was no longer my stream was where it entered the Byram River, a flow that during the times of Native American sovereignty was called the Armonck, or "fishing place," but which, according to one local legend, the English renamed because of the native tendency to pester white men with armfuls of shad and herring for trade and the endlessly repeated entreaty "Buy rum? Buy rum?" The Byram continued south for another ten miles after the juncture before widening and finally emptying out into the sea. The beginning of an idea came to me.

Several hundred dollars made it into my account after an ersatz bar mitzvah that my partially Jewish family cobbled together for me when I turned thirteen, and through a debt-leveraged matching grant from the depths of my mother's complicated finances I was able to purchase a used aluminum boat and a twenty-horsepower outboard engine. Using her good figure and her ability to forge solidarity with the working classes (she had been a friend of the American socialist Michael Harrington and was an experienced strike aide-de-camp), my mother persuaded the Greenwich harbormaster to let us jump the waiting list for a boat slip at the Grass Island Marina. By the summer of 1981, I had a boat, a place to store it, and several thousand square miles of sea for my own use. Better hunting grounds had been found at last.

This was not the time of child seats or swallowproof soda-can tabs. No safety seals secured Tylenol bottles or yogurt containers. Today it would be considered parental negligence, but in that first summer of boat ownership my mother would drive me down to the Grass Island Marina, seat belt–less, in her black secondhand luxury-edition Chrysler Cordoba and drop me off at my thirdhand boat. As I finished dumping my gear out of her trunk, she would light a Dunhill cigarette, cough heavily, and then, with a glance in her rearview, speed off into the childless afternoon ahead of her. So, at the age of thirteen, I learned how to navigate and fish on the sea by myself. It wasn't difficult—I'm sure most children, given the opportunity, could have figured it out. Once upon a time, being thirteen really did mean you were a man. But the feeling of steaming out into open water in pursuit of wild game, leaving the financial and physical constriction of mainland Connecticut behind, was exhilarating.

I did not have a GPS to plot my position or sonar to help me find fish. There was no cell phone to "check in" with home. I learned to find quarry by chasing flocks of diving terns or following a line of rocks from the shore with the assumption (usually right) that they indicated similar fish-holding rock piles down below the surface. If a rivet came loose from the hull of my boat—a sometime occurrence, since the hull had not been anodized to withstand salt water—I would slip off a flip-flop and hold the errant piece of metal in place with my big toe. I was sometimes able to persuade my older brother to join me, but midway through that first summer he announced that he "no longer wanted to kill things." I didn't mind. I was happy to be alone with the fish and the ocean.

By my second year of boat ownership, I began to understand the flow of fish as they came and left Long Island Sound. St. Patrick's Day, around when the forsythias first bloomed, was the time to test the mudflats for flounder just off the Indian Harbor Yacht Club docks. By April, when the forsythia shriveled to brown and the dogwoods came into flower, mackerel would have passed into the Sound and blackfish would be on the reefs surrounding Great Captain Island—a sure sign that it was time to put the boat in the water. Soon the lilacs would blossom, heralding the arrival of the first weakfish and porgies in May. And by the time lawns were being mowed with ferocity ashore, bluefish were coming into the harbors, devouring the mackerel and menhaden and everything else that had the misfortune to get in their way.

Striped bass, the prize of prizes, were also rumored to make an appearance around this time, though for me those fish remained only a rumor—they were already too rare for a poorly skilled captain to find them. By fall, blackfish would arrive again, along with a reappearance of flounder, and by wintertime, when my boat was back on blocks and nothing could be caught, I would enlist the much more substantial financial resources of my father and cajole him into taking me fishing on the *Viking Starship* party boat out of Montauk, where we would steam miles and miles offshore in search of codfish.

Before self-sufficiency became trendy and "locavorism" a catchword, I learned how to make my pastime "sustainable." Season by season I would take my surplus catch to the parking lot of my junior high and sell my fish out of the trunk of my mother's Cordoba for a dollar a pound. The miserably paid teachers would crowd around, and by the end of a sales session I would have enough cash on hand to buy gas for the next trip out.

The years of my boat and "my" ocean gave me a deep, atavistic belief in the resilience of nature. Even with the proximity of the Gatsbyesque mansions hugging the shorelines, the faint roar of I-95 audible as I cruised the bays, and all the other evidence of human civilization, Long Island Sound still felt to me like wilderness—a place to freely search out and capture wild game. I thought of the sea as a vessel of desires and mystery, a place of abundance I did not need to question. The ocean provides, therefore I fish. During my childhood I was often reminded how wealthy my neighbors and schoolmates were and how insecurely my family lived by comparison. The sea, meanwhile, was the great leveler. No fisherman, no matter how rich, had any more right than I did to a huge expanse of territory and resources.

But the desire to pursue fish and the desire to pursue females of your own species are inversely proportional. The fishing jones waxes from about age seven until sixteen or so and then abruptly withers in the harsh hormonal light of adolescence. Brief hot flashes of the fishing urge come on at times in the high teens, but they have an unanchored quality. The prime directive of life has shifted, and dusk no longer conjures the possibility of seabirds diving into a school of breaking game fish, but rather the moment when perfume and perspiration waft into the air, intermingled.

The summer I turned eighteen, my boat never left the two sawhorses it sat upon in wintertime, and it moldered, barnacle-covered, like a nautical version of Shel Silverstein's *The Giving Tree*, in the grassy parking spot adjacent to the last of my mother's rental cottages. And by the time I was nineteen and in college, I no longer came back to Long Island Sound. My mother sold my boat when I turned twenty. Fishing had done whatever good it was going to do for me as a man, she figured, and that was that.

But while erotic love between two parties tends to vanish for good when it exits, the bond between fishers and fishing has a way of circling back and restoring itself along different lines. After a decade and a half of various romantic false starts, often abroad, I found myself on the East Coast in my early thirties with a renewed desire to fish. Yet, like all mature loves, this next fishing phase aroused suspicion as well as pleasure.

The second fishing period of my life was also brought on by my mother. I had recently returned from working in Bosnia, a frustratingly landlocked and ruined place, where the best spots for natural idylls had been rolled over by tanks or scavenged to the bone by refugees. I often found myself staring down into the Drina River near the bullet-pocked city of Mostar, bluer than blue, but ultimately fishless from four years of war and subsistence angling.

During my time abroad, my mother and I had grown estranged and spoke rarely. This continued until she quickly ended the estrangement by receiving a diagnosis of metastatic lung cancer. I quit my job and spent the spring with her. Most of the afternoons of these three bad months were at her bedside, and, as can happen with someone fast approaching death, conversations were far-ranging, mundane, significant, and entirely out of context. Toward the middle of the second month, a clarity came over her; she could see the strain that her bad end was putting on me. One afternoon she sat up in bed and attempted to fix her wandering eyes into a focused, important stare. "Why don't you go fishing?" she asked, then fell back in a coughing fit.

Fishing? What a thought! But then again, why not? My brother was on hand to take care of our mother for the few days I'd be gone. It was April. A good fishing time for the East Coast, as I remembered it. Forsythias were still in bloom, and dogwoods were coming on, which meant flounder, blackfish, and mackerel. But when I called around to the tackle shops I had frequented in my youth, I found that the narrative of the spring migration had changed. The flounder season had been curtailed to only a few short weeks, and people spoke of a two-fish outing as a banner day, whereas once we had caught bucketfuls. Blackfish were hard to come by. Mackerel had not entered the eastern half of the sound in any numbers in a decade. A little bit farther afield, codfish that I used to catch with my father on divorced-dad weekends aboard the massive *Viking Starship* were almost nonexistent. There had to be fish out there somewhere, but the terrain had changed and I didn't know how to find my bearings. And when my mother finally died in June of 2000 and we spread her ashes at Tod's Point in Greenwich, the anglers who worked the shoreline there were fewer than I remembered and their buckets were generally empty.

Loss can have a tricky way of playing itself out in the mind of the loser. A psychologist once told me that in the face of loss either you can grieve the lost thing or you can incorporate it into your very being and thus forestall the grieving. Fishing somehow came to be that lost thing I clung to. With the help of my mother's 1989 Cadillac Brougham, a car she'd scarred badly on a stone pillar when the half dozen tumors in her brain had partially blinded her, I drove up and down the Connecticut and Long Island shores, northward to Massachusetts and Maine, and south again through the Carolinas and Florida, fishing all the way. And all the way, fisherman after fisherman echoed the same complaints: smaller fish, fewer of them, shorter fishing windows, holes in the annual itineraries of arrivals and departures, fewer species to catch.

In addition to fishing, I did one other thing that had been a habit of my former fishing self—I visited fish markets and tried to divine the provenance of what was on ice. The difference was palpable. If, after taking in a screening of the movie *Jaws* in the summer of 1975, you were to walk down to the bottom of Greenwich Avenue to the Bon Ton Fish Market near Railroad Avenue (as I did on many occasions), you would likely have found at least a dozen varieties of finfish displayed. Many of those fish would be from local waters. All of them would be wild-caught. They would buoy you up with their size and color, the clearness of their eyes, and the fresh quality of their skin.

But in the early 2000s, as I traveled the eastern seaboard, I saw that a distinctly different kind of fish market was taking shape. Abundance was still the rule, and yes, I still saw groupings of many species that could give the impression of variety and richness. But like anyone who fishes regularly, I have some ability for decoding the look of fish flesh, and I can usually tell how long ago a fish was caught and whether the names fish are sold under are quaint localisms or intentional obfuscations of something alien from far away. What I noticed was that in the center of the seafood section, whether I was in Palm Beach, Florida; Charleston, South Carolina; or Portland, Maine, four varieties of fish consistently appeared that had little to do with the waters adjacent to the fish market in question: salmon, sea bass, cod, and tuna.

Just as seeing my stream entering the Byram River had given me the idea to pursue the wider world of the ocean, seeing this peculiarly consistent flow of four fish from the different waters of the globe into the fish markets of America drew me again beyond the familiar to find out what had happened. I spent the next few years, sometimes on my own recognizance, sometimes for the *New York Times*, traveling to places I had previously only read about in the pages of *Field & Stream* and *Salt Water Sportsman*.

The more I examined the life cycles and the human exploitation of salmon, sea bass, cod, and tuna, the more I realized that my fishing history and the fishing history of humankind followed a similar pattern. Just as I had started out inland in a freshwater pond and then made my way down a river to coastal salt water when my grounds had gone bad, so, too, had early human fishers first overexploited their freshwater fish and then moved down the streams to their coasts to find more game. And just as I later turned to the resources of my father to take me far offshore to catch codfish beyond sight of land, so, too, had humans marshaled the resources of industry into building offshore fishing fleets when they found their near-shore waters incapable of bearing humankind's growing burden.

The more I thought of it, the more I realized that the four fish that are coming to dominate the modern seafood market are visible footprints, marking four discrete steps humanity has taken in its attempts to master the sea. Each fish is an archive of a particular, epochal shift. Salmon, a beautiful silvery animal with succulent pink flesh, is dependent upon clean, free-flowing freshwater rivers. It is representative of the first wave of human exploitation, the species that marks the point at which humans and fish first had large-scale environmental problems and where domestication had to be launched to head off extinction. Sea bass, a name applied to many fish but which increasingly refers to a single white, meaty-fleshed animal called the European sea bass, represents the near-shore shallow waters of our coasts, the place where Europeans first learned how to fish in the sea and where we also found ourselves outstripping the resources of nature and turning to an even more sophisticated form of domestication to maintain fish supplies. Cod, a white, flaky-fleshed animal that once congregated in astronomical numbers around the slopes of the continental shelves many miles offshore, heralded the era of industrial fishing, an era where mammoth factory ships were created to match cod's seemingly irrepressible abundance and turn its easily processed flesh into a cheap commoner's staple. And finally tuna, a family of lightning-fast, sometimes thousand-pound animals with red, steaklike flesh that frequent the distant deepwater zones beyond the continental slope. Some tuna cross the breadth of the oceans, and nearly all tuna species range across waters that belong to multiple nations or no nation at all. Tuna are thus stateless fish, difficult to regulate and subject to the last great gold rush of wild food—a sushi binge that is now pushing us into a realm of science-fiction-level fish-farming research and challenging us to reevaluate whether fish are at their root expendable seafood or wildlife desperately in need of our compassion.

Four fish, then. Or rather four archetypes of fish flesh, which humanity is trying to master in one way or another, either through the management of a wild system, through the domestication and farming of individual species, or through the outright substitution of one species for another.

This is not the first time humanity has glanced across the disorderly range of untamed nature and selected a handful of species to exploit and propagate. Out of all of the many mammals that roamed the earth before the last ice age, our forebears selected four cows, pigs, sheep, and goats—to be their principal meats. Out of all the many birds that darkened the primeval skies, humans chose four—chickens, turkeys, ducks, and geese—to be their poultry. But today, as we evaluate and parse fish in this next great selection and try to figure out which ones will be our principals, we find ourselves with a more complex set of decisions before us. Early man put very little thought into preserving his wild food. He was in the minority in nature, and the creatures he chose to domesticate for his table were a subset of a much greater, wilder whole. He had no idea of his destructive potential or of his abilities to remake the world.

Modern man is a different animal, one who is fully aware of his capability to skew the rules of nature in his favor. Up until the mid-twentieth century, humans tended to see their transformative abilities as not only positive but inevitable. Francis Galton, a leading Victorian intellectual, infamously known as the founder of eugenics but also a prolific writer on a wide range of subjects including animal domestication, wrote at the dawn of the industrialization of the world's food system, "It would appear that every wild animal has had its chance of being domesticated." Of the undomesticated animals left behind, Galton had this depressing prediction: "As civilization extends they are doomed to be gradually destroyed off the face of the earth as useless consumers of cultivated produce."

And that brings us to the present day, the crucial point at which we stand in our current relationship with the ocean. Must we eliminate all wildness from the sea and replace it with some kind of human controlled system, or can wildness be understood and managed well enough to keep humanity and the marine world in balance?

In spite of the impression given by numerous reports in the news media, wild fish still exist in great numbers. The wild harvest from the ocean is now around 90 million tons a year. The many cycles and subcycles that spin and generate food are still spinning, sometimes with great vigor, and they require absolutely no input from us in order to continue, other than restraint. In cases where grounds have been seemingly tapped out, ten years' rest has sometimes been enough to restore them to at least some of their former glory. World War II, while one of the most devastating periods in history for humans, might be called "The Great Reprieve" if history were written by fish. With mines and submarines ready to blow up any unsuspecting fishing vessel, much of the North Atlantic's depleted fishing grounds were left fallow and fish increased their numbers significantly.

But is modern man capable of *consciously* creating restraint without some outside force, like war? Is there some wiser incarnation of the hunter-gatherer that will compel us to truly conserve our wild food, or is humanity actually hardwired to eradicate the wild majority and then domesticate a tiny subset? Can we not resist the urge to remake a wild system, to redirect the energy flow of that system in a way that serves us?

In his landmark 1968 essay in the journal *Science*, "The Tragedy of the Commons," the ecologist Garret Hardin noted that "natural selection favors the forces of psychological denial. The individual benefits as an individual from his ability to deny the truth even though society as a whole, of which he is a part, suffers." What we have seen up until now, with both the exploitation of wild fish and the selection and propagation of domestic fish, is a wave of psychological denial of staggering scope. With wild fish we have chosen, time after time, to ignore the fundamental limits the laws of nature place on ecosystems and have consistently removed more fish than can be replaced by natural processes. When wild stocks become overexploited, we have turned to domestication. But the fish we have chosen to tame are by and large animals that satisfy whimsical gustatory predilections rather than the requirements of sound ecologically based husbandry. All these developments have gone on underwater and out of sight of the average modern seafood eater. We eat more fish every year, not just collectively but on a per capita basis, pausing only (and only briefly) when evidence surfaces of the risk of industrial contaminants in our seafood supply. Under the umbrella of these collective acts of denial, individual and corporate rights, national prejudices, and environmental activism have been cobbled together into something government officials like to call "ocean policy." In fact, there is no "ocean policy" as such, at least none that looks at wild and domesticated fish as two components of a common future.

But now, as wild and domesticated fish reach a point where they are nearly equal parts of the marketplace, this is just the kind of ocean policy we need. And in telling the story of four fish, for which the collision of wildness and domestication is particularly relevant, I shall attempt to separate human wants from global needs and propose the terms for an equitable and long-lasting peace between man and fish.

Salmon

The Selection of a King

If you were to go looking for a place where the problems between humans and fish first got serious, Turners Falls, Massachusetts, makes a worthy candidate. Located at a narrow pinch point halfway up the four-hundred-mile stem of the Connecticut River, Turners Falls is today the sort of hollowed-out New England former mill town that compels the traveler to move through quickly. Gloomy brick buildings line its main street, and the only encouragement to tarry is the public lot that charges just five cents for a parking spot.

But the most noticeable thing about the village of Turners Falls is that there are no falls.

There is only a dam several hundred feet across that metes out water in greedy spurts to the rocks below. No plaque commemorates the damming or explains why the river's progress was impeded in the first place. And there is no evidence whatsoever that before the dam the Connecticut River was an important salmon river, one of dozens of salmon rivers throughout New England and Atlantic Canada that made salmon an abundant wild staple for natives and early colonists alike.

Today in my native land of coastal Connecticut, there is no direct experience or memory of local wild salmon as food. The fish live in the minds of my fellow northeasterners as faceless orange slabs of supermarket product flown in from far away, eaten on bagels, and called "lox"—lox from the Indo-European *lakhs* and subsequently the Yiddish and Norwegian *laks,* meaning salmon. But salmon were once present here in significant numbers. The name Connecticut comes from the Algonquin *quonehtacut,* which translates as "long coastal river." For hundreds of years before my home state was a state, it was known principally as a place where a long coastal river wended its way to the sea and nurtured great annual runs of salmon, shad, and herring, an abundance that drew Native Americans from as far away as Ohio.

Every year perhaps as many as 100 million Connecticut River salmon larvae (no one knows exactly how many there were) would hatch out of large, bright-orange, nutrient-rich eggs. After spending one to three years in the fast currents of the river's tributaries, salmon juveniles (known as "smolts" at this phase) would pass over Turners Falls, heading down out of the mouth of the Connecticut. They would then shoot over to the fast-moving shunt of water in Long Island Sound called "the Race"—a treacherous spot where I once nearly overturned my small aluminum boat while fishing with a friend during summer vacation. Riding the Race's six-knot currents on an outgoing tide, the salmon would make a hundred-mile jaunt to Long Island Sound's terminus at Orient Point before breaking northeast twenty-five hundred miles to the Labrador Sea just west of Greenland. Upon arrival in Greenland waters, they would mix with other salmon from Northern Europe as well as with those from Spain. The Spanish salmon were in fact the first salmon, the strain that birthed the entire Atlantic salmon genome, which millions of years earlier had radiated out across the Atlantic. Though one might think a Spanish provenance would imply a warmth-loving animal, salmon originally hailed from the lush, cool valleys of Asurias and Cantabria in northern Spain and evolved to thrive in cold water. The colder the water, the higher the oxygen content, and salmon, with their hard-swimming, predatory metabolism, need a lot of oxygen. In Greenland they found not only cold, oxygen-rich water but also an abundance of oily krill, capelin, and other forage, which they consumed in large amounts and stored up as rich supplies of fats—fats that humans would come to associate with the heart-healthy omega-3 fatty acids, compounds that have the unique capacity to keep muscle and vascular tissue pliant and vibrant even in subzero temperatures.

Selection pressure in the form of seals, whales, disease, and accidents of various kinds culled away salmon throughout their journey, leaving less than 1 percent of the original hatchlings to complete their life cycle. After a sojourn of usually two years in Greenland, the survivors would go their separate ways, the American fish to the Connecticut's mouth at Old Saybrook and to many other rivers in New England and Canada, the Europeans to the rivers Tyne and Thames in England as well as rivers in Spain, Scotland, Ireland, France, Germany, and Scanadinavia on east into Russia. By the time they reached their home rivers, the salmon were big fish—broad-shouldered fifteen- to thirty-pounders with olive-silver backs and shimmering white bellies. Their flesh was thick and orange from the reddish pigment in the krill they ate and zebra-striped with enough energy-storing fat to propel them head-on against ten-knot currents. For reasons not completely understood, salmon do not eat upon return to fresh water and so must store great amounts of fat in advance of their spawning runs. These reserves made them great battlers on the line, so much so that when the seventeenth-century cleric–turned–fishing writer Izaak Walton was looking for a metaphor to hide his monarchical sympathies during the repressive Cromwell years, he called salmon "The King of Fish." This kingly impression extended to the table; special mention of salmon as royal table fare has been made by Roman and Scottish lords alike.

There were no lords waiting for the Connecticut River salmon when they returned to precolonial North America, though. Just native spearfishermen and netters, none of whom had any devastating effect on salmon numbers. The fish were more or less free to complete their genetic missions. Some had evolved to stop early on and spawn in the tributaries near the mouth of the river. Others were designed to sprint up Turners Falls and spawn all the way in the tiny rivulets that feed into the Connecticut from the Green Mountains and the White Mountains of Vermont and New Hampshire. The genome, the sum of the genetic components of the Connecticut salmon, was so broad that sub- and sub-subpopulations were able to make use of radically different tributaries, spawning throughout almost the entirety of the Connecticut's four-hundred-mile length.

During the colonial era, different chunks of the Connecticut salmon run were wiped out as millers dammed tributary after tributary for local power generation. But in 1798 a final death blow was struck. That year at Turners Falls, Massachusetts, entrepreneurs put a much larger dam across the main stem of the Connecticut. The salmon that had left for Greenland before the Turners Falls dam was constructed returned to find that they could not reach their spawning grounds. By the turn of the century, those old breeders had died off without ever getting a chance to reproduce. The broad, complex genetic potential of the Connecticut River salmon had vanished from the face of the earth.

Many salmon extirpations are more recent. It is possible that my generation (I am forty-two as of this writing) may be the last one to have a direct memory of wild Atlantic salmon at all. As recently as my early childhood in the late 1960s, Nova Scotia salmon, often called "Nova lox" by New Yorkers, were wild fish, harvested from several wild runs that spawned in the rivers of Atlantic Canada. But in the 1950s, after a handful of Danish and Faroe Islands fishermen found the patch of water off Greenland where all the world's wild Atlantic salmon congregated, they began catching tons of them. When Norwegian and Swedish fishermen joined the Danes and the Farose in the 1960s, wild Atlantic salmon went into perilous decline. Today a mere wisp of the wild Nova Scotia salmon population re-mains, and none of it is commercially fished. In fact, every appearance of the species *Salmo salar,* or "Atlantic salmon," in supermarkets today, be they labeled Canadian, Irish, Scotch, Chilean, or Norwegian, is farmed. Except for isolated pockets in far northern latitudes, there is no longer a popular memory of "wild Atlantic salmon" as food.

The Pacific species of salmon—the kings, cohos, sockeyes, pinks, and chums of the separate scientific genus *Oncorhynchus*—are another story. Those fish migrate from Russian and Pacific North-west rivers and use the Bering Sea as their Greenland and still reach the supermarket mostly from wild sources. But those wild salmon have also been winking out steadily in the course of my lifetime. There are diminished runs of them still remaining in California, Oregon, Washington, and British Columbia, but their viability is in question. California closed its salmon fishery completely in 2008 for the first time in history, and the famed Columbia River that divides Washington State from Oregon now hosts less than a tenth of its historical run of 10 million to 16 million fish.

So when it comes to salmon, the modern experience is a paradoxical mix of two phenomena. At one pole is the contemporary seafood counter, blooming like some kind of irrepressible orange rose, overflowing with fresh farmed Atlantic salmon fillets. These salmon are grown in monocultures as uniform and calculated as any animal feedlot and are the product of some of the earliest experiments in modern aquaculture. Because they lay large, oily eggs, visible to the naked eye, salmon are far easier to spawn and raise in captivity than many other common food fish, which lay small, nearly microscopic eggs. The first recorded experience of human-controlled reproduction of Atlantic salmon occurred in France around the year 1400, and since then salmon domestication has carried this single species clear across the equator to Chile, New Zealand, and South Africa—an entire hemisphere where, prior to the introductions of mankind, they had been entirely absent. The aquaculture companies operating in the frigid fjords of southern Chile now produce almost as much salmon per year as all the world's wild salmon rivers combined.

At the other pole of the salmon experience is the vanishing tail of wildness. In their Atlantic range, salmon have declined drastically throughout most of Europe, New England, and Atlantic Canada. In the Pacific the half dozen species and hundreds of genetically distinct strains of wild salmon are slipping away, river by river. What is left to us now are the two last primeval salmon territories: the wilds of eastern Russia and the forty-ninth American state of Alaska.

. . .

In the summer of 2007, an Alaskan fish trader named Jac Gadwill invited me to come visit him at the height of the king-salmon run on the Yukon River—the longest salmon river in the world. "Do be prepared for a bit of 'culture shock' here," Jac wrote. "Wonderful, loving people, but this is the USA's own third world country. The most remote, ignored area of the United States, with the highest unemployment and poverty levels. Fortunately it also has the finest salmon by far in the world. This is why the Yupiks (meaning is 'Real People') settled here over 10,000 years ago. We just yesterday shipped kings from here to some of New York's finer restaurants, direct to them, via FedEx."

Two weeks later, after swooping over the mountains that separate southern Alaska from its wilder northern part and then cruising in low over the Yukon River basin, I stepped out of a tiny propeller plane and entered the corrugated metal shed that serves as the airport in Emmonak, Alaska. A figure whose look could fairly be summarized as "a great bear of a man" stood squinting at me. There was something familiar about him—a kind of Nick Nolte of the North with a little more warmth and girth.

"You Paul?" Jac Gadwill asked, his voice thick with the grit of ten thousand packs of cigarettes.

"Yeah."

"Got here okay, did you?"

"Yep."

He took a pause, stared down at the floor for a moment, and then looked up and appraised me with his head cocked at an angle. "Boy, you look good here, Paul," he said finally. "You should stay."

We went out to an industrial-size pickup truck loaded down with fishing gear flown in from Anchorage, four hundred miles away. We headed along the gray ooze of a road that led through the clammy late-spring fog. On the way Jac had this to say about his thirty years in the Alaska salmon business:

"In the lower forty-eight, people are sort of arranged. They know when they get out of school what they're gonna do, what they're gonna achieve. In Alaska it's all mixed up. It's like everybody's running even along a mud track. But then all of a sudden, someone throws sand under one guy's feet and zoom! Off he goes. And you're like, 'How'd he do that?' Well, I'm kind of that person. A few years back, someone threw some sand under my feet, and off I went."

Soon we arrived at a bunkhouse and an adjoining office building. Both structures were tidy and snug-looking and semi-sunk in that same gray mud. A sign hung in front of the office:

KWIK'PAK FISHING COMPANY

NEQSUKEGCIKINA

"I asked an elder in a village upriver a ways what the Yupik word for 'good fishing' was, and that's what he came up with," Jac said. "Well, when I had that sign done, I showed it to the Yupik here in Emmo to see what they thought of it. They just kind of stared and said, 'Something to do with fishing, right?' Turns out the dialect's different in every village." Following this (and many other things he would say over the course of the next few days), Jac let out a smoky *"Wha-ha-ha-ha-ha"*—a raucous guffaw that brought to mind a cartoon character on the verge of launching a grandiose, doomed plan. "I tell you," he said, recovering from his laughing/coughing episode, "this thing is gonna kill me."

"This thing," as Jac likes to call the recently founded Kwik'pak Fisheries, is something both very new and extremely old in the ten-thousand-year interaction of man and salmon. What makes it old is its basic principle—native people in small boats fishing for wild salmon. What makes it new is the same principle—native people in small boats fishing for wild salmon. But unlike those who run the large canneries and fishing operations to the south in Alaska with a transient army of seasonal laborers, Kwik'pak's founders are working toward something different. Instead of white men coming along catching all the fish or ruining the fishery with a dam as they did at Turners Falls, this time the Native Americans will reap the profits. This time they will sustainably harvest the fish, brand it with a hyperlocal name, and sell it back to the white man at a premium. Kwik'pak is the only seafood company in the world that has earned recognition from the Fair Trade Federation for its labor and compensation practices. It is native-owned and largely native-operated, with the exception of a few outside managers and salespeople like Jac Gadwill. If all goes well, the Yupik board of directors of Kwik'pak hope that these particular native people catching these particular Yukon king salmon will bring a product to market that will be one of the most valuable fish on earth. How and why this is a possibility is the modern history of salmon itself, a history that is unfolding even as I write these words.

"Why don't you go take a look around town," Jac told me as he headed up the ad hoc staircase to his office. "I'm gonna go call Fish and Game and see if we can't get us an opening. I'll try the sugar-and-honey approach. If that doesn't work, I'll get my Lithuanian blood up."

. . .

Aside from the Kwik'pak Fisheries, an Ace Hardware store, and the local division of the state Department of Fish and Game, there is pretty much nowhere for the residents of Emmonak, Alaska, to go. Nor is there anywhere outside of town to go. Alaska is split at a diagonal roughly seventy-thirty between the northwest and the southeast. The southeasterly 30 percent has roads, outlet stores, McDonald's franchises, nail salons, psychiatrists, Californians' summerhouses, and a phone number you can call if you'd like to claim a moose that you saw killed on the highway. The northwesterly 70 percent of Alaska has very little of all that. Seen from above, Emmonak is very clearly in the middle of that emptiness—a gray divot dug up from a massive golf course of hundred-mile-long neon-green moss fairways and water hazards bigger than cities. No roads connect it to anything. And yet walking down the village's abbreviated thoroughfare, you can't get away from the traffic. Grandmothers in babushka scarves, fathers with sons riding piggyback, even children clearly under the legal driving age all cruise their all-terrain vehicles up and down the hillocks in the road, shouting in the cold, foggy air.

The Yupik nation barely noticed me as they zoomed around town. A woman in the distance called out enigmatically, "Sweetie, Sweetie!" A ways down, in the yard of a kind of jigsaw-puzzle house made of salvaged sky blue plywood, a man grasped the eye socket of a bloody walrus head with his left hand and sawed away at a tusk with his right. "Sweetie, Sweetie!" the voice called. A purebred pug appeared out of the fog and sprinted toward the voice. From the second story of another jigsaw-puzzle house, a man scolded, "You sleep all day! Good-for-nothing—you can't even catch fish! Damn Eskimo!" And for those who *can* catch fish, a yellow sign posted throughout the town by the local Fish and Game Department declared:

> **From June 1—July 15 a person may not possess king salmon taken for subsistence use unless both tips (lobes) of the tail fin have been removed. Clipping must be done before the person conceals the salmon from plain view.**

On this day the salmon situation was making the Yupik nation particularly idle. Everyone was waiting for the handful of white men and women at the Department of Fish and Game at the far end of town to determine if enough salmon had escaped into the upper river to allow for a commercial "opening" of the fishery. Every year in every major river system in Alaska, Fish and Game sets what they call "escapement goals,"—that is, a total quantity of salmon that must escape capture so that a sufficiently large number of adults make it to their spawning beds to lay enough eggs to ensure a viable next generation. When I arrived in Emmonak, the Department of Fish and Game was in a "conservative regime." They had been rattled since 2000–2001, when the Yukon king-salmon returns dropped far below their 53,000-fish average for still-unknown reasons. The fish's numbers had been slowly inching their way back up again, but the year of my visit, escapement goals were not being met, and Fish and Game was proceeding with a degree of caution that was making people like Jac Gadwill exasperated. Jac mentioned that he had heard rumors of death threats.

But seen in the greater context of what has happened with salmon around the world, it's easy to understand Fish and Game's caution. When it comes to salmon, Alaska is a little like a wise old man sitting on a far northern perch overlooking the destruction that humanity has wrought farther south. Almost visibly, the shock wave from the global near eradication of wild salmon seems written into the landscape of this richest of seafood states.

Before the Industrial Revolution, the world's population of wild salmon was likely to have been four or five times greater than it is today. Even in areas where there was no direct outlet to the sea, "landlocked" varieties of salmon evolved and used large lakes, like Lake Ontario, as their own private oceans. It is not for metaphorical reasons that the principal river draining into Lake Ontario from New York State is called the Salmon River. Nearly every river in Northern Europe, including the Thames and the Rhine, also teemed with them. The oft-told story of prisoners rioting on account of being served too many lobster suppers in colonial New England applies to salmon dinners and Scottish prisoners as well.

But salmon abundance requires a set of river characteristics that have stood in direct opposition to human industrial development, and salmon were among the first fish to suffer extreme extirpation at the hands of humans. Salmon need rivers that are free-flowing, clean and oxygen-rich, and protected by significant timber cover. One by one, each of these characteristics has been removed from the world's major salmon rivers. Free-flowing water has been eliminated first by small milldams and later by large hydropower complexes. Clean, oxygenated water has been voided by agricultural runoff and industrial effluent. Timber cover has been robbed outright by logging. And though these factors were well established and well known to be key to salmon survival since the 1800s, wild salmon as a commodity have never been economically valuable enough to deter the more immediately profitable human activities that destroy salmon. A remarkable memo from Julius Krug, the secretary of the U.S. Department of the Interior under Harry Truman, basically admits this.

"The overall benefits to the Pacific Northwest from a thoroughgoing [hydroelectric] development of the Snake and Columbia Rivers," Krug wrote in the 1940s on the eve of the construction of the dams that decimated salmon runs 16 million strong, "are such that the present salmon runs must be sacrificed." Only in retrospect and in the face of steep declines do humans smack their foreheads in dumbfounded realization and reach out, Lorax-like, for the last vestiges of wild salmon slipping from their outstretched hands.

In the late 1980s when I left college for a while, thinking I might disappear into the West and work as a fisheries biologist, I participated in one of these attempts at salmon salvage in rural Oregon. There, in the Willamette River Basin, I took habitat inventory of tree cover, built current diverters to create slack water for salmon juveniles, and trudged up and down streams all day long with a lazy career fisheries bureaucrat who listed on his employee self-evaluation that his greatest fear in life was falling into a river and drowning. We were looking for signs of spawning spring king salmon, a relative of the Yukon king and a fish that had lived in the Willamette River Valley for millennia. It was said to be one of the more delicious strains of the species. In my three months of stream surveying, I sighted one fish. To date there are very few examples of successful salmon restoration, for a variety of reasons. Often failures stem from the resistance of regulators to remove dams or restore streamside forests, but the vanishing or depletion of the original genetic material of the specific salmon run in question makes all restorations something of an uphill battle.

On an evolutionary scale, though, salmon have withstood epic cataclysms before—indeed, salmon species' exceptionally broad stock of genes buffers them against periodic and dramatic contractions of population and range. In the 50 million years of salmons' existence on earth, lava flows, ice ages, and the rearranging of mountains have wiped out thousands of miles of salmon territory on a regular basis. But after each contraction, the richness of salmons' genetic material has allowed populations to opportunistically seize on new habitat when it emerges. What makes the contemporary man-made salmon crisis unique and alarming is the effect humanity is having on the genome of all salmon species, simultaneously, throughout their global ranges. Pacific salmon are now extinct in 40 percent of the rivers where they were known to exist in California, Oregon, Washington, and Idaho and highly diminished in the runs that remain. In the whole of the Atlantic Ocean, wild salmon populations hover somewhere around five hundred thousand individuals compared with what may once have been a population of tens if not hundreds of millions of fish. It is in the wake of the salmon destruction carried out in the past centuries that fisheries' managers in Alaska have zeroed in on the maintenance of genetic diversity as one of the most important factors in preserving wild salmon. But they have had to do this in the face of ever-increasing demand. In the last three decades, the harvest of Alaska salmon more than doubled, to over 200 million animals annually.

But even with demand growing yearly, managers reserve the right to act conservatively when they think things are going in the wrong direction. This was why the people on Emmonak's main street, the people who fish the Yukon, had nothing to do. Salmon enter the Yukon Delta in bursts, and each burst represents a slightly different genetic subpopulation. After years of watching salmon runs implode, fisheries managers have learned that maintaining diversity within a given population is critical. Each burst may be headed for a slightly different bit of the Yukon's nearly two-thousand-mile-long watershed, and Fish and Game makes the argument that the more these sub-subpopulations survive and thrive, the richer the overall salmon genome is and the more adaptable and elastic the population will remain in the event of a crisis.

At the same time, Fish and Game has to make allowances for another population living on the river: Yupik Eskimos. Fisheries managers will permit "subsistence openings" for a limited number of hours, during which time the Yupik can catch salmon for their personal consumption. These fish have to be readily identifiable as subsistence catch and not for sale (hence all those yellow signs talking about the clipping of tail lobes). Only once the number of salmon in the river exceeds both the escapement and subsistence goals does Fish and Game allow a "commercial opening." And when a commercial opening takes place, the Yupik can sell what they've caught to Kwik'pak Fisheries.

On the Yukon a commercial salmon opening occurs in a relatively civilized fashion. There are only two fishing companies working the area, and the tribal unity of the people makes it basically a collaborative effort. In the more populous salmon regions to the south, where lower-forty-eighters often run the show, the moment Fish and Game declares an opening a dangerous game of waterborne, motorized rugby begins. Fish and Game draws a line of passage for salmon with floating buoys in the river, beyond which boats are not allowed to fish. Dozens of boats crowd the lines, bumping up against one another. Some boats are jet-powered, with no descending propeller, and can skip over other fishers' nets. As the day progresses, Fish and Game gradually reduces the fishing area. There is a crush as the managers draw in the line. If you cross that line, you can receive an initial fifteen-hundred-dollar fine. If you do it multiple times, you get points on your fishing license, a bit like drunk driving. If it goes on too many times, they take your license and your boat.

But even though this kind of wild competition does not generally occur, the shifting regulations still make things tense. When I finished my tour of Emmonak and returned to the Kwik'pak offices, Jac Gadwill shushed me with a finger while he listened nervously to the announcement over the radio. A woman with a flat midwestern accent droned out the bad news:

"At this time Fish and Game will not be opening the commercial king salmon fishery. There will be a *subsistence opening only* in the Y-1 and Y-2 section of the river from twelve to six P.M."

Jac slumped in his chair. He pulled a long drag off a cigarette and exhaled with a smoky cough.

"No milk and cookies for Fish and Game."

He took off his baseball cap and ran his hands though his unwashed, slightly-too-long, sandy-gray hair. He glanced over to the wall where a chart favorably compared the Yukon king salmon's fat content to that of other Alaskan salmon. Finally he pushed a button on the intercom and called out to his secretary.

"Hi, Jac," she said.

"Yeah, hi," Jac replied. "Can you see if Ray and Francine are around? I want to get Paul here out on the river."

Jac loaned me a set of orange rubber overalls and a thick, very comfortable pair of wool socks and wished me good fishing.

An hour so later, a Yupik Eskimo named Ray Waska Jr. threw the throttle all the way forward on his 150-horsepower engine, and his tiny metal skiff hurtled down the channels of the Yukon Delta. Francine, his wife, sat next to him in a camouflage outfit, and their teenage son, Rudy, perched at the front of the boat. Their three-year-old daughter, Kaylie, in racing-style pink sunglasses and a matching pink jacket, crouched between Francine's knees at the bottom of the boat. Their five other children were at the grandparents' fish-smoking camp, hidden away in the channels twenty to thirty miles upriver.

If e. e. cummings had wished to retire to a place where the world was truly "mud-luscious" and "puddle wonderful," then the Yukon River floodplain would have made a good choice. Minnesota boasts on its state license plate of being the Land of Ten Thousand Lakes. Alaska has 3 million, and it seems that a good number of them are the potholes and broken-off oxbows that surround the Yukon, the greatest of Alaskan rivers—a kind of Mississippi of the Arctic that bisects the state and continues far into Canada. There is so much of everything natural here—sky, wind, water, and, most memorably, clouds of insects that make a stinging helmet around your head the second the boat slows down.

How the Yupik find their way amid this shifting matrix of green sluices and bald shoreline is any white man's guess. Hardly a tree or rock marks the route, and as with any truly productive salmon delta, land is semipermanent, sinking or rising at the whim of the river. Yet there was never a hesitation in Ray Waska's steering. Turns were made with unquestionable assurance, until the engine cut out abruptly and Rudy Waska rushed to the front of the boat and started paying out net line, hand over fist. Suddenly we were subsistence fishing.

Once we set up, there was nothing to do. The net hung vertically in the water, a surface-to-bottom curtain a few dozen yards long blocking passage in a small portion of the river. There were so many salmon in the river at that point that even a partial obstruction in the current would result in fish. We were fishing with gill nets that had mesh openings big enough to accommodate the head and shoulders of a chum salmon—a less illustrious fish than a king salmon and sometimes called a "dog salmon."

The buoys strung along the top of the set net started to twitch. I had seen only one wild salmon in my life—that single fish I sighted in my fish-counting days in Oregon two decades ago—and I rose in my seat with excitement. But on the Yukon, even though this year was turning out to be a poor one, there were still several hundred thousand king, chum, and coho salmon expected to arrive throughout the summer. Ray and Rudy Waska barely noticed the salmon slowly filling their net, twitching the buoys. The rarer kings have heads that are bigger than the day's allowed mesh size, and they would be able to bounce off unharmed if they hit the net. It was all chums today. While chums are perfectly good to eat and also very sleek, beautiful animals, they are smaller, much more common, less fatty, and thus less prized by both Yupiks and nonnatives. Kwik'pak has recently been trying to rebrand chum salmon as "keta"—the native name—but the fish has yet to catch on. Nobody was in a hurry to haul.

But haul we finally did. After just four hand-over-fist pulls on the nets, the first three salmon were in the boat.

"Chums," Ray said, pronouncing the last consonants hard and sharp, the way that the Yupik tend to do with English words, making it come out as "chumps." We hauled some more and fish after fish flopped in the boat, their mouths and gills ripped up by the nylon net. The big white plastic well, about the size of a concert grand piano, in the center of the boat quickly filled up with salmon. It was a little like factory work. Haul, haul, salmon, salmon, flop, flop. But just as things started to seem commonplace, Ray tensed up. He pushed his son out of the way and expertly handled the net. He made one last haul, and *thwap!*—a much bigger, more beautiful salmon lay on the deck. It had accidentally snared itself in a net meant for chums, the twine wrapped thrice around its jaw.

"King," said Ray, the faintest trace of excitement in his voice.

The fish was about thirty pounds, twice as big as the chums, and had a steel-colored head that stood out from the rest of its body like a knight's helmet over chain mail. If the fish had not opened its mouth when it approached the net, it would not have snared its jaw. It would have bounced off and slipped through and advanced perhaps all the way to White Horse, Canada, where it might have laid its eggs and lived a fulfilled life. But instead Ray reached in and ripped out two of its gill arches, and blood poured onto the deck. A bled fish dies faster, and its value is increased because it lasts longer frozen.

Since Fish and Game had declared a subsistence opening only, the king salmon could not be sold to Kwik'pak Fisheries. But nobody had said anything about barter, something I supposed fit loosely into the category of "subsistence." When the grand-piano fish well was full to the brim with salmon, we pulled up anchor and blasted our way farther upriver. The wind was starting to penetrate my rubber overalls. The only parts of my body that were warm were my feet, stowed snugly in Jac Gadwill's socks.

Around a bend our boat slowed again. The insect helmet formed over each of us, and suddenly, rising up from the water, was a black oil tanker. It was making the long haul, taking oil out of the area of Alaska that is nowhere and transporting it to somewhere. We pulled up next to the ship and banged on the hull. Some prior communication had evidently taken place, because a few moments later a dude appeared on deck carrying two ten-pound packages of frozen chicken parts. Francine Waska stood and smiled and took the packages and laid them on the deck of the boat. They were an ugly reminder of the way the world is going. Yellow foam backing. Plastic wrap. A bar-code sticker that said "$19.99." Francine appraised the packages.

"Gee," she said, "I hope this doesn't have freezer burn."

Ray nodded to the galley cook and reached down into a cooler. With one huge haul, he grabbed the king salmon and threw it up onto the ship's deck, where it landed, shimmering beautifully, steel-colored in the watery sunlight.

A pause.

"Holy shit," said the cook. He looked down at it and shuffled his feet and glanced at the frozen chicken he'd traded in return.

"Hold on a sec." He slipped a hand into the gill plate of the salmon, dropped the fish, picked it up again, and disappeared into the galley. He returned in a moment with two more Safeway packages of frozen ground beef.

"Gee, thanks," said Francine. She looked at them and turned to me. "Do you think these have freezer burn?"

Before I had time to answer, Ray had loosened the rope and pushed his skiff back and once again we screamed down the river.

The Yupik don't seem to hold many grudges. Even after many centuries of unfair trading with the rest of the world, these kinds of exchanges are made with a minimum of reflection. Perhaps it's because the Yupik see the wild raw materials so plentifully within their grasp as essentially mysterious. The processes by which the world synthesizes sun, water, and earth into a slab of endlessly useful pink, healthful salmon flesh are unquantifiable. What is important is that those pink slabs return each year, uninterrupted, in large enough numbers to fill the Yupik smokehouses and drying racks so that folks can make it through the winter or sell enough to educate their children and improve a community that suffers one of the highest suicide rates in the United States.

The Fair Trade Certification of the Kwik'pak Fisheries is an attempt to try to mend the relationship of native fishermen with the rest of the world. A high price is sought for the Kwik'pak catch, and much of the profits from the company go back into the community. But no matter how much I nodded in agreement when told of the good intentions behind this new kind of fair fish trading in the world, I could not get out of my mind the more basic trade that I had witnessed aboard Ray Waska's skiff—the exchange of thirty-odd pounds of frozen, processed chicken and beef for a thirty-pound fresh king salmon from the wild currents of the Yukon.

The root of what seemed to me to be a quintessentially unfair trade stems from a more profound imbalance in the world. Whereas Alaskan salmon outnumber Alaskan humans by a ratio of fifteen hundred to one, the global human population outnumbers the global wild salmon population probably somewhere on the order of seven to one. If wild salmon were really the only option for the rest of the world to eat, then by all rights Ray Waska's king should have cost a fortune, exponentially more than that ground chuck and those chicken parts. But unlike the Yupik Eskimo mentality, the Judeo-Christian mind is governed by a faith in improvement and transformation of the natural world. The Yupiks wait for game to arrive. Judeo-Christians see the arrival of food on their plates as something that can be scheduled and augmented by focusing effort.

As early as the time of Moses, God commanded humans to seek out, select, and breed animals and plants in a way that would differentiate them from the wild melee around them. "Thou shalt not let thy cattle breed with unlike animals," God commanded Moses in one of the first published recommendations for controlled food culture. "Thou shalt not sow thy field with two kinds of seed." It is a commandment to isolate and focus our attention on a discrete set of plants and animals. To dewild them from their context, so to speak, and to grow them in an efficient monoculture.

Over the last four thousand years, this dewilding of animals has been accomplished primarily through a practice that has come to be called "selective breeding." From the time of Moses until the Industrial Revolution, we have progressively selected individuals within animal populations possessing sets of traits that suit our purposes. This "improvement" of our livestock occurred slowly at first, with animals becoming gradually more useful decade by decade. The slowness of the progress was due mostly to the fact that when humans first began selecting traits, they selected them according to what they could see. It was understood since the Roman era that a white-faced cow would have a good chance of producing another white-faced cow. A speedy sire and a quick dam were seen as good bets to create another fast horse. An ignorance of the unseen genetic truths that lay behind these traits kept humankind from delving any deeper.

This breeding by outward observation was encapsulated by the first truly systematic animal breeder, the British animal-husbandry pioneer Robert Bakewell. In the mid-eighteenth century, Bakewell coined the phrase "like begets like" and set about isolating sheep and cattle that had traits he felt were universally appealing to breeders. So confident and relentless was Bakewell in his breeding practices that he created entire family lines of sheep and cattle that still form the basis of the world's major animal breeds. The "like begets like" school of thought continued into the early twentieth century with an ever-increasing degree of complexity, but it took a gut-sucking world depression for the next step in animal breeding to emerge.

At the height of the Great Depression, a professor at Iowa State

College named Jay Laurence Lush began codifying the *internal* traits
of animals into a system of breeding that selected not what individual animals looked like but rather how efficient an entire population of animals could be at turning feed into flesh. The child of
farmers, Lush was forever preoccupied with the practical. No doubt
he had observed through breeding on his family's own farm that a
"like begets like" approach had certain limitations. As he grew to
adulthood and was increasingly surrounded by hungry countrymen
who could not afford the price of meat, he began looking into how
traits could be systematically and more accurately passed on to subsequent generations.

Throughout the 1930s and '40s, Lush developed a collection of
theories that distilled down to their basic elements could be summarized as this: Improving just one animal is not enough to bring
about rapid change in the productivity of farm animals. The true
expression of progress that we seek is the improvement of a whole
population, a new race, if you will. Instead of trying to breed one ideal
animal, breeders need to focus on moving the average qualities of an
entire population closer to an average that is more in line with what
humans can use.

And more than anything else, what humans could use out of a
population of animals was more meat for less cost. In animal husbandry, feed is traditionally the biggest cost for any farmer. Before
Lush and his theories were applied, many animals required as much
as ten pounds of feed for every pound of meat they produced. But
over time, by coming to an understanding of the genetics that regulate growth within a population at large, breeders were able to apply
Lush's principles and accelerate growth rates so that the "feed conversion ratio"—that is, the number of pounds of feed required to
produce one pound of meat—could be lowered substantially. It is this

accomplishment that enabled the galley cook of that Alaskan oil tanker to buy ten pounds of chicken parts from Safeway for the astronomically low price of $19.99. The animal that produced that meat came to market twice as fast after consuming only half as much feed as an animal that Robert Bakewell would have raised.

Though the work of Lush continued in terrestrial animals, there was one major limiting factor that slowed the rate of improvement of a population over time: cattle and sheep produce only a few offspring in the course of their lives. The progress of discovering which parents create the most productive animals was limited by the small sample size of each new generation. Many crosses of many families were dead ends further limiting progress. Much backtracking had to be done. Improvement, relatively speaking, was gradual.

But in 1963 a meeting between Jay Laurence Lush and a young Norwegian animal breeder named Trygve Gjedrem suddenly opened up an entirely new avenue. For the Yupik nation and anyone else in the world who had anything to do with wild salmon, that meeting would change everything.

Trygve Gjedrem is semiretired now, but you can still find him animatedly moving around the offices of a Norwegian research institution called Akvaforsk. Akvaforsk's offices are located in the town of Ås, nearly as far north as Kwik'pak Fisheries but on the opposite end of the human/salmon relationship.

To get to Akvaforsk, you must first pass through the IKEA-showroom-looking Oslo Airport and then travel south for half an hour on a local train, yellow and clean and as steady on the rails as a zipper. Unlike most other European or American cities, Oslo gives up quickly to the countryside, and within a few minutes the whitest of snows blankets the pleasantly rolling hills, dairy farms, and cozy-looking wooded hamlets. Crisp, well-defined cross-country-ski tracks run alongside the train, and Norwegians, who seem more comfortable on skis than they do on foot, whisk by in precise, healthy strokes, sometimes keeping pace with the train as they glide downhill.

Perhaps it was the snowy northern climate where I met Gjedrem, but sitting there in a little leather cap with blue twinkly eyes, he looked to me like one of Santa's more senior elves. When it comes to salmon, it turns out, he is much more like Santa himself.

If he had proceeded along with life as he originally intended, Trygve Gjedrem would have had nothing to do with salmon. He was trained as a sheep breeder, and sheep were what he knew best. During his youth Gjedrem and most of the rest of the European agricultural community were captivated by the success that Americans were achieving in improving animals for human consumption. This was part of a larger trend in the agriculture of the 1960s that came to be known as the "Green Revolution"—a series of scientific leaps in crop and animal development that caused agriculture to become substantially more productive. The Green Revolution is largely credited with having successfully staved off famine in India, China, and elsewhere in the developing world just as populations were booming. And in 1963, when Gjedrem went to the States as part of a foreign-exchange program, he was thrilled to meet one of the principal architects of the animal side of the Green Revolution, the animal-breeding theorist Jay Laurence Lush. "Lush was a fantastic man," Gjedrem told me as the snow sparkled outside his window, "a great man. But he was a quiet person. He did not use hard words."

Unbeknownst to Lush, there was an experiment going on in Norway at the time of Gjedrem's U.S. sojourn that would greatly amplify the influence of his theories. Beginning in the early 1960s, around the same time as wild Atlantic salmon were being fished into oblivion off the coast of Greenland, two brothers in the Norwegian town of Hitra named Sivert and Ove Grøntvedt began collecting salmon juveniles and raising them in nets suspended in the clear waters of the local fjord. Of all fish, salmon proved particularly adaptable to this process. Generally speaking, most of the fish we like to eat hatch out of microscopic eggs and require microscopic food to get through the first phases of life—something very hard to replicate in an artificial environment. Salmon, however, hatch out of large, nutrient-rich eggs and live off an oily yolk sac for the first weeks of their lives. They are quickly able to transition to eating chopped-up pieces of fish. Something the Hitra brothers were able to obtain easily from the dense herring population in the fjords of coastal Norway.

The Hitra trials overcame an essential problem that happens with salmon in nature. With most salmon a substantial number of young die in the early phases of life. This mortality may be more than 99 percent in natural systems. But by keeping the fish protected from predators in net cages and giving them a regular food supply of herring and other small fish, the first salmon aquaculturists reversed nature's equation. Suddenly many more animals were surviving, and with wild salmon already in steep decline those fish could be sold at a considerable profit. "They really earned money!" Gjedrem told me, slamming the table with his open hand on each downbeat. "And they told their brothers and sisters around the coast, 'WE MADE MONEY!'"

Seeing the success of the Grøntvedt brothers, Gjedrem and his thesis adviser, Harald Skjervold, realized that the breeding logic of Jay Laurence Lush, if applied to salmon, had huge potential. Up until the meeting with Lush, the initial profits being made in the nascent Norwegian salmon-farming industry were being gleaned from fish that were essentially wild in their genetic makeup. No one had done the hard work with salmon breeding that Lush and his four thousand years of predecessors had done with cattle and sheep. "I am a breeder," Gjedrem told me, "and we thought it was important to get started by first selecting a breed of fish. If there was going to be real success, we realized we could not have efficient production based on wild animals."

Moreover, the Norwegian breeders had one thing that modern cattle breeders didn't have: a vast genetic reservoir of wild animals from which to draw the most favorable genes. Since wild cattle were domesticated many millennia ago, without any coherent genetically based selection methodology, many useful genes may have been lost and never made it into the animals we eat today. But at the time Norwegian salmon breeding began, wild salmon were still viable and diverse. The genetic potential was enormous.

The initial selection of farmed Atlantic salmon took place from fish drawn from forty different river systems. Every salmon river has its own unique set of challenges to which fish must adapt. Some rivers are very long, like the Yukon, and require animals that can build up tremendous fat reserves in order to survive the extended journey. Others are very far north, with only a short season of warmer temperatures, and require a fish that can maximize growth, particularly during its juvenile phase. But whatever the manifestation of difference that occurred in different strains of salmon, the first salmon breeders realized that crossing and recrossing the specific families from the original forty rivers would result in salmon that grew faster. And because salmon, unlike cattle and sheep, can produce *many thousands* of offspring in the course of their lives, once favorable individuals were found, just a few matriarchs and patriarchs could form the basis of a whole new race of highly productive fish. A domestic population could be created quickly that would be quite different from the initial wild forebears.

For Gjedrem and the other breeders of Akvaforsk, it was as if they had discovered a new continent of possibility. "The goal with growth rate is to get upstairs," Gjedrem told me, sketching a rudimentary staircase on a piece of paper in front of him, "This footstep—that's the generation interval. And the game is to step up. Because of Lush's theories, we were sure that we could walk up the stairway with salmon. The first results showed us that there were dramatic differences between the best growing families and the worst. . . . And what is so impressive is that each generation, each step up, we made progress of thirteen- to fourteen-percent improvement in growth rate."

In other words, within just seven generations—fourteen years—the Norwegians were able to double the growth rate of salmon—something that had taken thirty generations and sixty years of applied breeding, not to mention an unknowable amount of Neolithic-era undocumented selection, with cattle and sheep. The end result was the breeding of a fish that while still technically the same species as its forebears was markedly different in its internal metabolism. Some scientists refer to this separate line of salmon as *Salmo domesticus*. By the standard of sheer numbers, *Salmo domesticus* is now the most successful salmon in the world. For it was *domesticus* that the Norwegians were to use when they turned salmon farming from a domestic endeavor to an international juggernaut.

The emergence of *Salmo domesticus* helped Norwegians increase production of farmed salmon to a world-dominating half million tons in just thirty years. Once the Norwegian fjords were full of salmon cages, the farming methodology and the genetic stock of *domesticus* were exported by Norwegian salmon companies to other cold-water, fjord-rich territories like southern Chile, Nova Scotia, and British Columbia. Indeed, the most striking thing about Chile's largest fish market in the Patagonian town of Puerto Montt isn't the exotic kingklip and the fist-size barnacles on display. It is the five-foot-high piles of bright orange salmon fillets shining slick and fresh in the austral sun.

Before the Norwegians came along, there were no salmon living in the world south of the equator—the equator acts as a thermal barrier that the cold-water-requiring wild salmon could not cross in nature. Today there are hundreds of millions of salmon in Chile, which is now the second-largest salmon-producing nation in the world. A further result of Gjedrem's efforts is the outright domination of farmed salmon over wild salmon. Every year more than 3 *billion* pounds of farmed salmon are produced, around three times the amount of wild fish harvested. Many of those many millions of farmed salmon, whether living in Norway, Chile, or Canada, can trace their heritage back to the breeding lines created at Akvaforsk in 1971.

To people who trade in wild salmon, like Jac Gadwill of Kwik'pak Fisheries, this seems like the worst kind of bastardization. "A cage is a cage is a cage," Jac told me when I asked his opinion of farmed salmon. "The life of a wild animal is completely different to the life of an animal in a feedlot. What happens to a fish if you don't let it swim? I suppose you could take a Fijian boy and raise him in Guyana and maybe he'd still wind up a fat boy, but I don't know."

But Trygve Gjedrem sees nothing wrong with a dominant strain of domesticated fish emerging in the world. Indeed, there is something in artificial selection that needs to be kept in mind when thinking about the health of the ocean in larger terms. Farmed salmon are the most consumed farmed finfish in the Western world. The salmon-farming industry requires an enormous amount of food. And with salmon a lot of that food consists of other fish that are harvested from the wild. In an unimproved state, farmed salmon require as much as six pounds of wild fish, ground up and turned into pellet feed to produce one pound of edible flesh. Selectively bred salmon, meanwhile, have reached a point where less than three pounds of wild fish can produce a pound of salmon. And as salmon continue to be bred into a more and more efficient consumer of marine protein, that ratio is likely to drop.

But there is also a risk. The tamed-salmon genome is now markedly different from the wild-salmon genome. When tamed salmon escape into the wild (as they do in the millions every year) they risk displacing a self-sustaining wild fish population with a do-mesticated race that is not capable of surviving without human sup-port. *Salmo domesticus* has been bred to eat a lot and grow fast in a controlled environment, but it has lost many of the fierce, deter-mined traits that make a wild salmon able to swim against powerful currents, withstand fluctuations in temperature, and spawn in a river besieged by predators. Critics argue that escaped farmed salmon may outcompete wild salmon in some phases in their life cycle only to be unable to reproduce later on down the line. Some maintain that this could have a fatal impact on the long-term viability of wild salmon everywhere.

In spite of these risks, Gjedrem believes that improvement should be the norm for all farmed fish. "With the exception of At-lantic salmon, we are so far behind terrestrial food production," he told me, driving me in his little blue car back to the little yellow train across the snowy white Norwegian dales. "Think of the Green Revolution of the 1960s! Since the Green Revolution, there has been no major starvation in India or China. The same thing should have started by now with fish and shellfish."

Of all the people I've met in the world of seafood, Gjedrem seemed the most baffled by the way salmon farming has been increasingly targeted by nonprofits as a polluting, environmentally degrading industry. Gjedrem is a child of the Depression, and the formative experiences of his childhood were poverty and human starvation. Any move away from that baseline is progress. His blue eyes twinkled, and he seemed to bristle with excitement when he talked about all the people the ocean could feed if breeding principles were put into place in a rational manner. "It's such a waste of resources," he declared of the world's failure to embrace selective breeding of fish. It was not in fury or anger that he said this, but with a kind of bewilderment. Why even allow for the possibility of starvation?

As we reached the train station and said our good-byes, I remembered one last thing I'd meant to get his opinion on. I told him how I'd heard that farmed salmon descending from the original Norwegian breeding lines had escaped from their net pens in Canada, and there was evidence that they were establishing themselves in west coast rivers. At this, Gjedrem smiled and smacked the steering wheel of his little blue car.

"Hah!" he said. "I was wondering when that was going to happen."

There was neither concern nor criticism in his voice. Just the quiet observation of someone of an earlier generation. Someone who saw the interplay of wildness and domestication as an ongoing drama where mankind was the central character and human starvation the archest of enemies.

. . .

Starvation is a phenomenon still very much alive in the memories of the Yupik nation, particularly the memories of tribal elders. True, the younger generation has grown up accustomed to having access to frozen packages of chicken parts and ground chuck from the lower forty-eight, but grandparents still recall a time when the only thing that got you through the winter was salmon.

A day after our processed meat–for–wild salmon swap on the Yukon River, Ray Waska drove me two hours upriver to his family's fish camp. There Laurie Waska, the seventy-five-year-old matriarch of the Waska clan, put me to work breaking down four hundred pounds of salmon. The camp consisted of a tidy blue house at the center of a clearing, a corrugated-steel smoking shed, and a four-legged corrugated-steel canopy under which Laurie and I sat. Dozens of grandchildren, some directly related, some adopted, ran around in the grass and mud.

Using a fan-shaped *ulaaq*—a fish-cutting knife—fashioned from the blade of an old circular saw, Laurie got to work on the salmon. Opposite from the way a commercial fish cutter would work, she started her filleting at the bottom of the animal, making a slit on either side of its anal fin and then hewing the meat upward toward the top. The fillets were smooth, orange, and flawless. If subsistence fishing doesn't pan out, Laurie could probably make a good living behind the counter at Zabar's or any other premium New York retail salmon outfit.

When I gave it a go, I was extremely conscious of her staring at me. In this subsistence environment, I was trying to fillet as close to the bone as possible. Laurie frowned at what I had done and took the *ulaaq* away from me.

"Too much meat," she said.

"I was trying not to waste."

"Too much meat."

I tried to do another salmon following her instructions, angling the *ulaaq* up as I cut to make a fillet about an inch and a half thick. A thinner fillet, it turns out, is better for smoking and drying. It is moisture that ultimately causes rot, and a thinner cut will allow water to work its way out of the flesh. Laurie picked up another *ulaaq*, and we worked silently in tandem. She did three salmon for every one of mine.

"That one's pretty good, isn't it?" I said, holding up my second fish.

"It'll dry." She stared down at the pink-orange mess of meat and bones that accumulated at our feet. We were literally up to our ankles in lox.

The inherent seasonality of wild salmon, the handful of weeks of extreme salmon abundance followed by months and months of no salmon at all, is a problem with which both Native American subsistence fishermen and Western salmon entrepreneurs have always had to contend. The Yupik address the problem by building smoking and drying sheds. Nonnative Alaskans, however, dealt with the problem by putting salmon into a can.

Before salmon farming was invented, most people did not have access to fresh salmon. Pollution and dams had ruined any salmon river that was unfortunate enough to be near a large human population center. Industrialized human societies and wild salmon have, with very few exceptions, never found a way to live harmoniously in proximity to one another. And so in the prefarming days, the only way wild salmon could reach the majority of consumers was in a can from Alaska.

To this day the majority of the salmon infrastructure in Alaska revolves around canning. You can see this in any of the small towns up and down the coast of the state. Over the course of the last century, entire factories were built at the mouths of rivers with huge vacuum tubes extending from their roofs down into the holds of waiting tender boats that in turn gather up salmon from the smaller skiffs working the river. From the wildest of provenances, the fish are converted into sliced orange chunks and amalgamated together on palates of unlabeled "bright stacks" at the backs of the riverside factories. They are differentiated only when an order comes in from one of the big canning marketers, at which point they are reincarnated as Bumble Bee, Icy Point, or Ocean Beauty.

The only choice that middle-class homemakers had for years was canned wild salmon, baked by our grandmothers into all sorts of horrendous casseroles and croquettes. Farmed salmon changed all that. Unlike canned salmon, which may sit on shelves for years at a time, most farmed salmon comes to rest on ice before our eyes at the seafood counter within forty-eight hours of its death. Moreover, unlike wild salmon, which traditionally came to market only during specific seasons usually only a few months long, farmed salmon is available fresh year-round. And as the Norwegian (and later Chilean and Canadian) breeders increased the feed efficiency of farmed salmon, the price became lower and lower—so low that today it is comparable with the price the oil tanker's galley cook paid for the ground chuck he traded with Ray Waska out on the Yukon.

But there was a strain of people who most decidedly did not like farmed salmon. Many of farmed salmon's detractors were keepers of the vestigial recollection of wild salmon that was slipping away from human memory. People who had sportfished for salmon, perhaps, or those who during the 1970s environmental movements had become familiar with Native American folktales of the wild salmon runs that had been lost. There were also owners of coastal property, particularly in Maine and Washington State, who did not like the look and the smell of salmon farming that began creeping up the coasts of Canada and the northern U.S. coastal states in the late 1980s and early '90s.

Fish farming in its first incarnations is almost always a privatization of a public resource—a mad-dash grab for ocean farming sites that previously belonged to no one. And the more efficient salmon farming became, the more environmentally problematic the industry became. The increased efficiency of improved, selectively bred salmon caused the fish to flood onto the market. Prices plunged. Farmers desperately opted to expand and grow more total pounds to compensate for the loss in per-pound revenue. Good farming sites with strong currents and clean water became rare. Farms were sited with poor water circulation and often in proximity to passageways for dwindling runs of wild salmon. As density of salmon farms increased, nitrogen wastes built up, causing algae to bloom and die and, in the process, deoxygenate the water. Overcrowding of farms attracted parasites, like a bloodsucking creature called a sea louse, which has been shown to be transferable from farmed populations to wild salmon runs. Diseases like infectious salmon anemia were born, first in Chile and then in the rest of the world, wiping out whole farms in a week. Diseases and pollution are classic problems associated with any kind of animal husbandry, but in the case of salmon farming all of this occurred within the context of a wild environment. And above and beyond all that, there was the essential feed equation that to many environmentalists didn't make sense: why use three pounds of wild fish as feed in order to generate just one pound of farmed salmon?

But while all these problems were significant, each one was hard to quantify. No one quite knew how many wild salmon were suffering as a result of farming operations. No one quite knew how much waste was building up in coastal waterways. No one knew if the continued harvest of wild "forage" fish for salmon feed would do long-term damage to marine ecosystems. Aquaculture facilities when viewed from land are innocuous-looking, a daisy chain of a dozen or so hoops with nets hanging below, floating in a flat plain nearly even with the water. There is a fishy smell at feeding time, and water can grow cloudy in extreme concentrations, but to the untrained eye the effects seem minimal. Nonprofits and coastal advocates flailed their arms and tried to get public attention, but no one seemed to take notice. In the early 2000s, however, a different approach was launched that drew on the experience of earlier food reform movements. Looking back on the commercial success of his 1906 meat-packing-industry exposé *The Jungle*, the best-selling Socialist author Upton Sinclair once lamented, "I aimed at the public's heart and by accident I hit it in the stomach." Taking a page from Sinclair, marine conservationists realized that the way they could bring attention to the problems with the salmon industry was to aim for the public's stomach directly.

David Carpenter is a gentle-eyed, white-haired physician whose offices are in a Legoland-style spur off the main campus of the University at Albany, smack in the shadow of General Electric, viewed by environmentalists as one of the most serious polluters of New York State's Hudson River.

Carpenter's training is in medicine and public health, but over the years his research has focused on toxicology and his advice has been sought out in relation to polychlorinated biphenyls, or PCBs, a by-product of the manufacture of electric insulators, flame retardants, and, most recently, computer chips. Over the course of the twentieth century, General Electric's plant on the Mid-Hudson River had discharged over a million pounds of PCBs into the river. In the 1960s it was discovered that those PCBs had entered the aquatic food chain and passed on into wild fish. Striped bass, a fish that spawns in the Hudson, was one of the first indicator species for PCB pollution and was largely responsible for the U.S. government's lowering the PCB contamination threshold on fish from five parts per billion to two parts per billion. But PCB contamination has spread beyond the rivers into the ocean at large. "What we are seeing is the overall contamination of the oceans and the food web within the oceans," Carpenter wrote me recently. "The rivers have contaminated the oceans, and the PCBs are getting bioconcentrated within the ocean food web."

Carpenter had on numerous occasions testified about the ill effects of PCBs, which include liver enlargement, memory loss, and fetus mutations. Most devastatingly of all, Carpenter's research showed that PCBs were a tremendously difficult compound to flush from the human body. The same inertness that made PCBs an ideal flame retardant and insulator made them equally impervious to human enzymes that try to cleave and eliminate them from the body. "The average half-life of a quantity of PCB," Carpenter told me, "is ten years." In other words, ten years are required to remove half of a quantity of PCB contamination from the human body. A person who ingests a sizable quantity of PCBs in his teens will likely be carrying around at least some of the chemical until he dies. No PCBs have been manufactured in the United States since 1977, but their legacy lives on in the fatty tissue of Americans.

In 2002 the issue of PCB contamination began to draw the interest of a major U.S. foundation called the Pew Charitable Trusts. For the preceding decade, Pew staff had been tracking the environmental impact of salmon farming but were frustrated by the failure of the public to understand the scope of the problems associated with these operations. Pollution, the spread of disease and sea lice to wild populations, the genetic mixing of farmed and wild populations, the grinding up of wild fish into salmon feed—none of it seemed to grab the popular imagination. As Joshua Reichert, the managing director of Pew's Environment Group, told me, "The public as a whole doesn't care much about the problems associated with the farming of marine fish." None of these issues, Reichert said, "seemed to affect people in the way that they approached farmed salmon and wild salmon or their proclivity to buy one or the other." Even worse, Reichert felt, consumers seemed to perceive salmon farming as a net gain for the environment. "The public has been led to believe that the production of farmed fish actually lessens pressure on wild stocks," said Reichert, "and we did not believe that to be the case. In fact, we believe the opposite is true." It was consumers' lack of information about farmed fish and their general failure to understand the bigger ecological issues around domesticating salmon that drove Reichert and his staff to start seeing if farmed salmon had a connection to something consumers *did* care about: their own health.

Reichert and others at Pew had heard reports that several samples of farmed salmon had shown higher levels of PCBs than wild salmon. Based on these initial hints, they decided to commission the largest study ever undertaken of farmed and wild salmon, with Ronald Hites at the University of Indiana leading the research in conjunction with David Carpenter.

When Hites, Carpenter, and other members of the study examined the flesh of salmon from around the world, they found that there was an overall difference in PCB contamination between farmed salmon and wild salmon. This is not due to any kind of genetic engineering or because the water that farmed salmon swim in is in some way polluted; contamination in salmon comes from what salmon eat. PCB pollution occurs all over the world, particularly in the Northern Hemisphere. PCBs enter the food chain when microscopic plankton absorb the chemical across their cell membranes. Small fish then eat the plankton and, because PCBs are not easily flushed from body tissue, retain increasingly greater amounts of PCBs the more plankton they eat. When small fish are ground up into feed pellets for salmon, PCBs are again transferred further up the food chain. Just as the little fish "bioconcentrated" PCBs in their flesh when they ate plankton, salmon bioconcentrate PCBs even more when they eat small fish. Generally speaking, PCB concentrations are amplified with every step up the food chain.

But wild salmon, it turns out, eat differently from farmed salmon. Two species in particular—wild sockeye salmon and wild pink salmon—are practically filter feeders, subsisting on tiny shrimp and other small crustaceans. This near microdiet is one or more "trophic levels" below the fish-derived pellets that are typically fed to farmed salmon. Since PCBs and most industrial pollutants tend to amplify every step up the food chain, the lower on the food chain a salmon eats, the fewer contaminants a salmon is likely to have in its tissues. Added to this is the fact that PCBs tend to accumulate in fatty tissue. Farmed salmon average 15 percent fat content and wild salmon average around 6 percent, so wild pink and sockeye salmon have a bioconcentration of PCBs much closer to that of other filter feeders than to fish-only-eating farmed salmon. If you were to take those same wild Alaskan salmon out of their native environment, put them in a pen for a year, and feed them manufactured feed pellets from a contaminated source, their PCB levels would rise. Generally speaking, bad feed equals bad fish.

But within this cardinal rule, there are also more subtle corollaries. Feed differs from region to region. In Southern Hemisphere countries (like the world's second-largest salmon producer, Chile, for example), salmon are considerably cleaner. That is simply because, overall, there is far less industry in the Southern Hemisphere than in the Northern; therefore Southern Hemisphere feed pellets are correspondingly lower in industrial pollutants.

But the subtlety of this information did not make it into the press. Like all information on food safety, it reached the public in binary fashion—wild salmon are good and farmed salmon are bad. And there was an immediate drop across the board in farmed-salmon consumption. Perceptual improvisations also occurred in a kind of toxicological game of telephone. Armchair environmentalists have often pointed out to me that farmed salmon have high levels of mercury. In fact, mercury contamination in farmed salmon is not a particularly salient issue. No significant difference in mercury levels has been found between farmed and wild salmon, and neither farmed nor wild salmon have dangerously high levels of that contaminant.

The Hites and Carpenter study also spurred counterattacks from the salmon-farming industry, which claimed that there were important benefits from eating oily fish like farmed salmon that outweighed risks from PCBs. "Long" fatty-acid chains found in salmon, such as docosahexaenoic acid (DHA) and eicosapentaenoic acid (EPA)—often referred to collectively as the omega-3 fatty acids—are used by fish to keep their cell membranes pliable in cold-water environments like coastal Greenland and Alaska. When eaten by humans, these amino acids have the same effect on human vascular tissue—keeping veins and arteries fitter and more youthful longer. The salmon industry argued vehemently that this effect was not being taken into consideration by the Pew-funded study. This position was amplified when the National Institutes of Health funded a study by the Harvard Medical School's Dr. Dariush Mozaffarian. The Mozaffarian study compared the risks of cancer death from PCB poisoning related to farmed-salmon consumption with the risks of coronary heart disease death from *not* eating farmed salmon. When I spoke to Mozaffarian this past year, he told me he felt that comparing the PCB cancer risks of eating farmed salmon with the coronary risks of not eating oily fish like farmed salmon was like "comparing sesame seeds with watermelons." Mozaffarian's meta-analysis found that 23 cancer deaths per hundred thousand individuals were likely to occur if people ate three portions of farmed salmon per week. If the same hundred thousand people did *not* eat farmed salmon or other oily fish, 7,125 deaths from coronary heart disease were likely to occur. Carpenter and others have subsequently countered that they believe Mozaffarian's selection from the scientific literature for his meta-analysis was not representative of the larger trends and failed to take into account preliminary evidence that PCB contamination in farmed salmon may offset the coronary benefits of omega-3 fatty acids that an eater of farmed salmon would likely obtain.

Mozaffarian says he would and does feed farmed salmon to his two-year-old child. Carpenter maintains that farmed salmon is "dangerous food."

There is, however, one point on which Mozaffarian and Carpenter agree. A single 1.8-gram pill of omega-3 oil supplements, available in forms that are guaranteed PCB-free and harvested from sustainable sources, provides as much coronary benefit as eating salmon, farmed or wild.

The lower-forty-eighters and Europeans who eat the bulk of the world's farmed and wild salmon are endlessly obsessed with prolonging life and avoiding long-term health risks. But the native people who catch wild salmon today, like the Yupik nation of the Yukon Delta, seem to have an increasingly tenuous hold on existence. A bad turn of fate can be all it takes to make a tribal member voluntarily leave this troubled world, particularly if that tribal member is a young person. While I was out at the Waskas' salmon-smoking camp, a pickup truck in the Yupik town of St. Mary's drove off the road and killed two of the teenagers inside.

"Now there's gonna be eight suicides," Jac Gadwill said.

He looked as if he hadn't slept. It was unclear whether it was the deaths that had kept him up or the Arctic summer sun, pouring in the windows of the Kwik'pak Fisheries bunkhouse all night long.

"It's true," Jac said to me. "You'll see."

A little while later, we crammed ourselves aboard a single-engine airplane and took off into the fog. The Yupik nation is an archipelago of settlements strung up and down the width and breadth of the Yukon Delta—an area about the size of the state of Oregon. Some of the encampments disappear when the river shifts—a wily river that can freely slip its banks is always a sign of good salmon country. Other settlements grow into villages and begin the slow creep toward a kind of modernization. And though the different outposts are separated by vast amounts of space and time, crises are somehow shared. It therefore behooved Jac Gadwill to fly to St. Mary's and comfort the families of the victims, to help stem the loss in some way.

Jac's long legs doubled up against his body. The readout on the navigation system in the plane said NO USABLE POSITION. DEAD RECKONING ON. The moment the plane left the ground, Jac broke off our conversation and fell asleep on the shoulder of his companion, Chong Cha (Ci Ci for short). Ci Ci owns a chain of nail salons in Olympia, Washington, and has an elaborate white pattern painted on each of her very long fake fingernails. She was only up for a visit and seemed anxious to leave again. She prefers it when the fishing season is over and Jac takes his winter vacation from Kwik'pak and lives with her in Olympia.

En route to St. Mary's, we stopped in the village of Kotlik, one of Kwik'pak's satellite fishing stations. Because salmon are always on the move, Kwik'pak must maintain several different harvest and shipping operations—a logistical nightmare in country that barely has any infrastructure at all. Jac had just installed an electronic time-card system for the employees in Kotlik, and he was eager to get it up and running. The moment we landed, he awoke, and we sprinted out of the plane. We quickly hopped aboard two waiting ATVs and trundled along a warped, slick boardwalk. Soon we had pulled up to a new loading dock at the side of the river. Jac pointed to an older, dilapidated dock just next door.

"I built that dock out of scraps that floated down the river a few years ago. Got weathered in for three days while I was doing it. Slept on the floor of that trailer. No food."

He looked around and squinted toward a field in the distance. "Now we got it so we can fly a Herc in here and fly it out to Seattle with twenty thousand pounds of fresh salmon. Get it to New York in a day."

Back when Jac Gadwill first began working in the Alaskan seafood business, the idea that anyone in New York would want a fresh

Yukon king salmon on his or her plate would have seemed preposterous. The thought of building an airport to accommodate such a difficult logistical feat would have seemed outright crazy.

But thanks to the taste for fresh salmon that farmed salmon developed and also thanks to the fear of PCBs in farmed salmon that the Hites-Carpenter study propagated, there is today an ever-growing market for fresh wild salmon. A market large enough to make flying a Hercules C-130 transport plane to a remote dirt runway at the top of the world twice a week seem both reasonable and potentially profitable. Seven-odd years after the Hites-Carpenter study came out, the debate about PCBs and salmon has gradually faded into a gauzy haze, and farmed-salmon consumption, as well as salmon farming in general, which both dipped after the study was published, have continued to grow again. But so, too, has demand for fresh wild salmon. In fact, in the last twenty years, consumption of all salmon, farmed and wild, has doubled. And it is here that it seems the environmental community who would use fear of PCBs as a tool for raising awareness should take heed. PCB contamination is clearly an issue of concern and one that should be addressed. We should not be dumping dangerous chemicals into the ocean, into a place we depend upon for our food. But questions about how we should go about farming domesticated fish and how we should go about catching wild ones is another set of concerns that needs to be addressed directly. The real dilemma at hand for consumers is the original issue that motivated the commissioning of the Hites study in the first place: salmon farming is in dire need of reform, and wild salmon stocks are under intense pressure and severely reduced in their range and potential. Industry has recently figured out a way to strip PCBs from salmon feed, but no one has figured out a way to bring huge amounts of salmon, farmed or wild, to market that is sustainable over the long term.

For if we were to rely solely on wild salmon, we would inevitably come up against the essential imbalance that exists between the needs of people and the requirements of fish: the imbalance that happens when humans choose wild Yukon kings over farmed ground chuck. Because of the way that people have altered the environment, there really are very few rivers like the Yukon in the world. Rivers where the water is cool and rich in oxygen. Rivers that flow and shift relatively unimpeded, where dams don't block passage and large, valuable old trees fall into rivers and create slack water for salmon juveniles. And it's unlikely that there will be such rivers ever again. Humans now outnumber wild salmon by a ratio of seven to one. What would happen if every human on earth demanded wild salmon instead of farmed salmon? Instant extinction.

Except that Western civilization, with its imperative to select and improve, has for the moment tried to tweak the number of wild salmon. Though the Yukon remains a very pure wild-salmon domain, the Alaska Department of Fish and Game's operations now stock many millions of hatchery-raised fish into the state's more southerly rivers every year to "supplement" wild production. Today nearly one in three "wild" Alaska salmon begins its life in a hatchery. This was a trick that was tried in the salmon rivers of the lower forty-eight to very ill effect. In Washington, Oregon, and California, hatchery-bred salmon were often introduced into rivers to which they were not endemic. As the fading wild populations diminished, the introduced hatchery fish ended up displacing the wild spawners, wreaking havoc on the original population of fish. Soon those salmon rivers were on human life support. If humans stopped stocking fish in most western rivers of the lower forty-eight, the salmon would all but disappear.

In Alaska the picture is more complex and, according to Alaska

Fish and Game managers, more carefully thought through. Nowadays, Fish and Game officials told me, Alaskan hatchery supplementation is river-specific. Two- to three-inch baby salmon that are put into Alaskan rivers come from parents of that same river. The Copper River is stocked with fish whose parents originated from the Copper River. Diversity is maintained throughout Alaska, managers say, by preserving the specific genetic integrity of each stocked river.

But as one salmon farmer pointed out to me, the genetics of these "wild-stocked" fish are still heavily skewed. In nature one of the greatest selection pressures on wild salmon is the weeding-out that happens during the highly vulnerable transition from egg to juvenile fish. In a natural river system, free of any human manipulation, as many as 80 percent of all eggs die before hatching. By putting stocked fish in a wild system, fish that have already been artificially reared from egg to two-inch juvenile, fisheries managers are circumventing one of the greatest natural-selection pressures of all. They in effect let bad eggs live and grow to pass on their bad genes to future generations. Who knows if those hatchery-raised fish, when they mature and lay their eggs, will produce offspring capable of surviving the full cycle of the wild gauntlet?

Indeed, it's possible that many Alaskan salmon no longer contain the genetic wherewithal to endure. A large chunk of wild Alaskan salmon may already be on human life support without our really knowing it. The confounding thing is that there is no way of finding out if we've gone too far. The only way to learn whether the salmon we've stocked in Alaskan rivers can endure is to stop stocking them. If "wild" Alaskan runs then disappear afterward, it will mean that we've made a terrible mistake.

I couldn't help but think that in a way the future of wild salmon and the future of the Yupik people were somehow sadly parallel to each other. Without some kind of outside management—the many grants and loans and education incentives and tours to the lower forty-eight that the federal and state governments provide to the seven thousand remaining members of the Yupik nation—it's doubtful whether any native Alaskans would stay in the Yukon River floodplain, this land that is at once rapturously beautiful and staggeringly depressing. If we continue as we're going, Alaskan salmon may need similar support to stay viable. Yearly stocking may become as imperative for their survival as food stamps are to the Yupik.

The Cessna airplane carried Jac Gadwill and me farther toward St. Mary's. The families of the teenagers who had died in the overturned pickup truck awaited Jac's condolences. Besides being Kwik'pak's acting representative at the time, he is a big, comforting presence, and there seemed to be an atavistic need for him to attend the mourning. But the cloud ceiling had fallen considerably, and Jac peered through the windscreen of the airplane at the dull, empty landscape ahead.

"We ain't gonna make it," he said.

"We might make it," the much younger bush pilot said, dipping the plane down a couple hundred feet into a patch of smoother air.

Jac lowered his head and whispered in my ear with a barely audible rasp, "We ain't gonna make it."

Suddenly he looked weary. A tiredness seemed to seep out of his bones and into mine. For a while Jac had believed that he was in his mid- to late fifties, but the previous year at the Boston Seafood Show a colleague who'd known him for most of his many years as an Alaska fish trader added things up and pointed out to him that he was in fact sixty.

"Do you really need to do all this?" I asked him finally.

"I tell ya, Paul. I made a million dollars in a day once," Jac said offhandedly. "Other times I say I came to Alaska with six hundred dollars in my pocket and it's taken me twenty years to make back my six hundred dollars." Then a pause. "But no, I guess I did all right in the end. I guess I'm kind of doing it for this," he said, pointing at his heart. "See, I'm a Catholic, and I kinda think that people, when they end their life, should have done something worthwhile." He grew silent and looked out the window, then muttered, "But this thing is all-consuming."

"Are you grooming a successor?" I asked, fishing perhaps too aggressively for a good quote. Jac looked at me over the tops of his glasses, recognizing the switch in tone, demanding greater sincerity. "I mean," I rephrased, "ever think of retiring?"

"Ha!" he said, reaching across the aisle for the dragon-lady fingernails of his companion. "Ci Ci here is my retirement."

The cloud ceiling fell lower, and the undesirable gray mass of St. Mary's in the distance seemed suddenly unobtainable.

"I'm sorry, folks," the pilot said, "we aren't gonna make it."

"Knew it," Jac said under his breath.

The plane banked low and hard, and we headed back toward Emmonak. We sat in silence until the now familiar town came into view once again. "Well," Jac said, leaning forward and squinting out the window at a town where he knew nearly every person by name, "better to be here than halfway to there."

Back in the airport shed, I found that the airline schedule had been completely upended and my flight to Anchorage had left hours ago. But after a quick conversation and a wink from Jac, the dispatcher rerouted me on another plane going to Anchorage via the town of Bethel.

"You better watch yourself in Bethel," Jac said. "You take a wrong turn there and you're gonna be writing for the *Tundra Times*— *wha-ha-ha-ha-ha!*"

I gathered up my bags from Jac's parked truck and made my way toward the waiting airplane. Jac reached out and unexpectedly gave me a giant hug.

"Damn, Paul," he said, shaking his head, "you look good here. You could make the cut. I could mentor you."

I felt a strange flutter in my heart and laughed. I shook Jac's hand and climbed aboard the plane. We took off with a short lunge forward and an easy lift, as small planes tend to do—a safer, more human-size hop upward that the big jets lack. We passed over the jagged mountains that separate the Yukon floodplain from the civilization in the warmer south, and I could still feel the dank Emmonak air lingering in the cabin. The only parts of my body that were warm were my feet. I realized that one small piece of Emmonak and the Kwik'pak Fisheries had left with me. Removing my boots and letting my feet air out, I felt Jac's presence. I don't think I could ever fill his shoes. But I was happy to leave Emmonak with Jac Gadwill's socks.

Since I left the Yukon in 2007, the fortunes of wild and farmed salmon have diverged further and the curse of the continuing downward spiral of wild salmon seems to continue to nip at my heels. In 2008 and 2009 almost no king salmon returned from the sea. The Yupik nation now faces outright starvation and its region has been approved for federal disaster relief. Nevertheless, the Yupiks recently took up a collection and donated forty thousand dollars to the Red Cross to aid survivors of Hurricane Katrina.

As in the previous Yukon salmon dip in 2002, no one quite knows why the kings have stopped running up the Yukon. Nowadays when something goes wrong (or continues to go wrong) with a wild-salmon run and no problem can be detected in the run's home river, most biologists gesture in the direction of the sea in a vague sort of way and say that "something is happening in the ocean." Oftentimes this "something" is that most ancient of problems: fishermen catching too many fish. It is a problem that has already been identified with the much more dire declines in the populations of Atlantic salmon.

In some cases the problems happening in the ocean are finally being addressed. In Iceland a former herring fisherman named Orri Vigfússon has pioneered a project to buy out the few remaining commercial salmon-fishing operations in the Atlantic in order to halt the commercial fishing of all Atlantic salmon. From the Faroe Islands south to the Irish coast, he has helped remove nets through-out the North Atlantic; indeed, he imagines a day when a kind of international salmon reserve will be created in the oceans from Rus-sia's Kola Peninsula all the way to Labrador, Canada. Salmon bio-logists agree that Vigfússon's efforts are bearing substantial fruit. Adult salmon are not being caught and killed in the ocean and are returning to rivers in Iceland in numbers that haven't been seen in many years.

In the Pacific the unpredictable population shifts still confound fisheries biologists, and things tend to be summarized in the sen-tence "We don't really quite know what is happening." But there are suspicions. Jon Rowley, a seafood consultant who had up until 2007 been trying to market Yukon king salmon as a premium fish, believes it is the fault of the Alaska pollock industry.

The Alaska pollock industry is the largest wild fishery in the United States and a fishery that has been twice certified as "sustainable" by the world's foremost sustainable-seafood endorser, the Marine Stewardship Council (more on this later). Two years ago, over 120,000 king salmon were caught accidentally as "bycatch" in pollock nets; a third of these salmon were probably destined for the Yukon River. That amount is more king salmon than all that the Yupik manage to harvest in a good year, killed as accidental bycatch. By law these unintentionally caught fish must be dumped overboard, dead. When I asked Rowley if anything had been done to challenge the "sustainable" certification of Alaska pollock and the effect they are having on the Yupik, he answered that the chiefs were making their way to Anchorage to testify before the regional fisheries management council, but, he concluded, "there is intense lobbying by the pollock industry to do little or nothing." Climate change is likely also affecting the Yukon kings, but no one quite knows how to quantify such an epochal shift.

Meanwhile, at the same time as Yukon kings and other wild salmon are having greater and greater difficulty swimming through the various impediments humans have thrown in their paths, another kind of salmon is gradually slipping through a different kind of obstacle course. The farmed-fish breeding effort the Norwegians have undertaken has brought about many results, some of them good, many of them questionable. We have successfully selected for fish that can come to market having needed half as much food as their wild ancestors. But, taken too far, the endless quest for a more and more efficient animal ultimately leads us up a dubious alley—an alley that goes beyond selective breeding into the realm of outright genetic manipulation.

AquaBounty of Prince Edward Island, Canada, is today the leader in trying to make a more efficient salmon through DNA manipulation. According to AquaBounty's Ronald Stotish, the attempts to genetically engineer a faster-growing salmon began in the late 1980s, when researchers started to look at the antifreeze genes that allow fish to survive in subzero-temperature water. But, as Stotish wrote, "Once the research progressed, they also realized that these interesting proteins had other potential applications." The antifreeze-gene research looked promising for a number of different medical, food, and cosmetic uses, and that research was spun off into a separate enterprise.

But perhaps the most lucrative thing the initial antifreeze research pointed to was faster growth. Stotish continued, "We were also interested in exploring whether or not we could improve the growth rates and economics of growth for Atlantic salmon by adding a second copy of a salmon growth hormone gene." Since they had figured out how to turn the antifreeze gene on and off, they realized they could use those same "switches" in association with the salmon's growth hormone. A trial was run, and researchers witnessed spectacular increases.

With more research and development, AquaBounty was eventually able to create a salmon that grew twice as fast as the already double-growth speed of selectively bred salmon. The new fish, trademarked as AquAdvantage Salmon, was recently submitted to the U.S. Food and Drug Administration for approval. To date there is no genetically engineered salmon on the market, but there could be a few years hence.

Looking at the two examples, the wild Yukon king salmon on one side and the modified AquAdvantage Salmon on the other, it struck me that the human/salmon relationship has been polarized into two intense extremes, either of which could collapse under the weight of its own presumptions. On the one hand, there is Kwik'pak Fisheries, a pure and noble attempt to bring a pure and noble fish to the world market. But a wild salmon is a resource that is ultimately so limited and variable that any attempt to maintain it in a world market is a risky endeavor.

At their root the wild strains of salmon in Alaska have a very narrow threshold for exploitation, and their move from niche item to world commodity could lead to a classic fisheries collapse. If we are going to continue to eat wild salmon, we must eat them sparingly as the rarest of delicacies and their price should reflect their rarity in the world. Even though the pink and chum salmon of the Alaskan rivers farther south are of lower quality than the Yukon kings, one could argue that there is little logic in supplementing their numbers artificially so that they can be sold in supermarkets at two dollars a can.

At the other extreme is the headlong effort toward efficiency at any price. AquAdvantage, a salmon so efficient that it will require very little feed and will ultimately be extremely cheap, will be capable of grabbing a huge market share if consumers can ever get past their discomfort with genetically modified food. AquaBounty's Stotish says that the risk of genetic contamination is minimal. "Our product will be all female, and sterile (unable to reproduce)," Stotish wrote. "Furthermore, we have applied to grow the fish in physically contained production systems. Examples of this could be tanks, raceways, etc. that prevent the escape of the fish." This is somewhat in line with what a number of environmentalists advocate. "Closed-system" aquaculture, in which salmon are raised in tanks away from natural systems, is the only way to guarantee that wild and domesticated forms of salmon stay separate. But these systems are costly. A modeling exercise conducted in 1998 by a consulting firm in the Bay of Fundy found that the only closed-system, out-of-sea models that showed a profit after five years were those that grew transgenic salmon. If your goal is to grow the most salmon by using the least amount of feed, then logic dictates genetic manipulation to be the best avenue.

It would be wonderful if all the salmon we eat could be wild. But as one marine ecologist said to me recently, to continue to eat large wild fish at the rate we've been eating them we would need "four or five oceans" to support the current human population. Over the last two hundred years, by reducing the amount of habitat that can support salmon and, at the same time, fishing hard on the stocks that still do exist, we have been eating our way into a deficit situation. We are eating into the principal, so to speak, of salmon stocks instead of harvesting the annual "interest," which is what people like the Yupiks used to remove modestly when the salmon returned to the Yukon every year.

A solution, of course, would be for those who don't live in salmon country to stop eating salmon altogether and eat smaller fish that have a smaller overall footprint on the sea. The idea of good consumer choices as a driver of change in ocean policy has become a leitmotif for contemporary chroniclers of the ocean's crisis. In this vein one writer suggested in an opinion essay for the *New York Times* in 2008 that New Yorkers should dispense with lox and bagels and have sardines with their cream cheese instead.

But the salmon industry is now a multibillion-dollar business, active on every continent in the world. A dip in consumer demand occurred in the wake of the PCB scare but resumed shortly thereafter and continues to grow by the year. Indeed, salmon is now a keystone industry at the very core of the international food industry. As one salmon farmer told me, "Most supermarkets wouldn't even have a seafood section if it wasn't for salmon." The power of consumer choice is a pleasant notion, but it has so far motivated little change.

What seems to me more necessary is a move to reform the laws and practices that govern the salmon industry. Salmon aquaculture is still a very young endeavor, less than forty years old in most countries. It is not yet set in its ways, and it is not necessary that the worst practices of the past become the standard practices of the future. There is still a chance for incorporating all we have learned about the problems of terrestrial monocultures into the relatively new frontier of aquaculture.

In July near the end of my salmon research, I found the beginnings of this new way of thinking when I drove up the coast of Atlantic Canada to the town of St. George on the Bay of Fundy. It was there that I met Thierry Chopin, a cheery and optimistic French transplant to Canada who signs off his e-mails with an encouraging quote from Jules Verne: *Tout ce qui est impossible reste à accomplir*— All that is impossible remains to be accomplished.

Chopin works in conjunction with the largest fish farmer in Maritime Canada, Cooke Aquaculture, developing a practice called integrated multitrophic aquaculture, or IMTA. This method of farming combines species that require feed (such as salmon) with other species (such as seaweeds) that extract dissolved inorganic nutrients and species (such as mussels and sea urchins) that extract organic particulate matter, to provide a balanced ecosystem-management approach to aquaculture. Like Kwik'pak Fisheries, IMTA's basic concept is very old. The world's very first aquaculturists, the Chinese who farmed carp starting four thousand years ago, began as polyculturists. Early Chinese silk farmers found that carp would naturally congregate under the mulberry bushes where silkworms would spin their cocoons. Eventually it was discovered that carp could be a crop in and of themselves. This original two-way relationship expanded over time. Carp feces, it was found, would stimulate the growth of rice and other useful grasses, which the Chinese harvested. These grasses also fed ducks that could be slaughtered for meat. Thus a four-way polyculture developed, with silk, fish, fowl, and grain all coming out of the shared and multiply repurposed fertility of a single pond.

When modern-day salmon aquaculture was launched in the 1960s and '70s, the concept of polyculture for some reason got lost. Early farmers were so thrilled by the prospect of bringing a high-value species to market for very little money that feedlot-style monocultures quickly sprouted up in some of the most pristine salmon country along temperate coasts around the world. Little attention was given to the siting of farms, the effects of effluent, or the spread of disease. In time, places like the Bay of Fundy became practically open salmon sewers, where effluent was released unchecked, cloaking the bottom with the ooze of salmon refuse.

After facing a series of crises and opposition from environmentalists throughout the late '90s and early 2000s, the industry began to restructure itself. In 1996 there were early signs of the presence of infectious salmon anemia in New Brunswick. This caused the New Brunswick provincial government and the industry to develop and implement, in 2005, a system of bay management areas (BMAs) that more carefully allot salmon sites. The move reduced the density of fish per site, introduced biosecurity measures, and required portions of the Bay of Fundy to be left fallow on a regular basis.

All these changes in the aquaculture industry also opened up the door for Dr. Chopin, a seaweed expert who had been doing research on kelp in Atlantic Canada since 1989, when he moved from France to the University of New Brunswick–Saint John. Seaweed, it turns out, is an integral part of the food, cosmetics, and textiles industries and constitutes a $6.2 billion market. Chopin had been working on the production of carrageenans, the thickening or emulsifying agents extracted from red algae that are particularly useful to industry. In an "aha" moment Chopin saw that the inorganic waste from salmon farms could be used to grow those very valuable algae species.

"Coming here to Atlantic Canada, I realized, 'Wow, with all this salmon aquaculture, we have all these nutrients in the water,'" Chopin told me as we motored out to one of Cooke Aquaculture's IMTA sites. "Instead of wasting these nutrients, why not recapture them?" Chopin recognized that larger organic particulate waste would also have to be dealt with. Collaborating with Dr. Shawn Robinson, from the St. Andrews Biological Station of Fisheries and Oceans Canada, he discovered that mussels could recapture midsize waste particles suspended in the water column. Later they found that they could also add organisms feeding on the heaviest particles of all—the ones that fell to the bottom. Valuable sea urchins and sea cucumbers, it turns out, are particularly fond of this kind of waste.

Still, IMTA is very much a pilot project. Chopin and Robinson started their collaboration with two smaller salmon-farming ventures, one of which was Heritage Salmon. When Glenn Cooke, the CEO of Cooke Aquaculture, acquired Heritage Salmon in 2005, he decided to scale it up. The polyculture experiments are still only a tiny part of Cooke's overall footprint, but they are expanding.

As we left the circular salmon pens and motored past the rectangular rafts of seaweed, Chopin drew my attention to a series of cages supporting hanging socks of blue mussels. Grabbing a mussel and opening it with a knife, he pointed to the delicate shimmering meat inside—it was spread out almost to the edge of the shell. "You can see here, it has almost thirty percent more meat than mussels that are typically available in grocery stores. And the nutritional profile is very favorable, too. There are significant quantities of omega-3 fatty acids, particularly the heart-healthy ones, EPA and DHA." Mussels turn out to do another interesting thing on a salmon farm. Evidence suggests that they may absorb some of the infectious salmon anemia virus; adding mussels to the aquaculture equation could serve to break the disease cycle that is rife in some of these salmon-farming operations.

None of the polyculture species can do anything about sea lice, perhaps the most pernicious effect of salmon farming. Nevertheless, there did seem to me to be a better future, one where "feed-conversion ratio" would not be simply a matter of pounds of feed going in to pounds of salmon going out. Rather what would result would be an array of seafood products in a cycle. Even Chopin, who has a love of graphs and charts and PowerPoint presentations, can't quite get a handle on how much food could be generated from such an operation. "In the chart the arrows are going everywhere, and I just can't calculate it yet," he told me.

Finally, IMTA could lay the groundwork for the elusive "closed circle," the quest of quests for sustainable seafood producers, one where the inputs and the outputs emerge from a single unit, with *zero* feed having to go into the system. This may not be as far off as we think. As Rick Barrows, an experimental-feed developer for the USDA, explained to me, "Fish require nutrients, not ingredients." It turns out that the nutrients, particularly the omega-3 fatty acids, present in the oft-criticized wild-fish feeds can be duplicated by seaweeds. The omega-3 fatty acids that occur naturally in salmon ultimately derive from seaweeds that smaller fish ingest before being eaten by salmon.

In a sophisticated polyculture environment, salmon would bypass the smaller fish that eat seaweed and would eat feed pellets synthesized from seaweed directly. By feeding in this way, we would in effect be reducing the trophic level of farmed salmon, turning them from predators into something closer to filter feeders. This would result in fish markedly lower in PCBs than those animals fed with unpredictable wild-fish feed sources. And the beauty of the system is compounded by the fact that the waste those salmon generated would in turn feed mussels and also grow more seaweed. Fish meal and oil would still be needed as very early feed for juveniles and to maintain the health of broodstock fish, but these would be minimal compared to what is needed at present in a traditional salmon monoculture.

Some purists argue that this is a bastardization of a salmon. That a salmon is naturally a predator and should naturally eat fish. An oft-quoted trope of the anti-salmon-farming camp is that we shouldn't be farming the tigers of the sea." But as Rick Barrows at USDA pointed out, this is a question of perspective. "We *can* farm the tigers of the sea," he told me, "as long as we feed them hay."

The unavoidable truth is that way back in the Middle Ages, when the first attempts were made at domesticating salmon, we should have chosen something else. There were most definitely better, more efficient fish out there. But we simply didn't have the technology to tame those other fish. Salmon's large eggs, their responsiveness to human intervention, and a lot of applied breeding science has advanced the human/salmon relationship to a level of complexity not seen with other marine animals. Quite simply, we *know* the salmon better than most other fish on earth. We have mapped large portions of its genome, crossed its families, and studied its life cycle intimately. To start anew with a completely different animal at this point would mean many decades of backtracking.

And so we've reached a crossroads with salmon. Either we can invest money and effort into making a more and more artificial salmon, one whose very genetic components are profoundly different from their ancestors, or we can simply say that we've gone far enough with selective breeding. That the selection that should happen now is the means of feed and husbandry practices that sustain these farmed fish. Instead of putting artificial selection pressure on salmon, it may be time to put selection pressure on *farmers*. Let the fittest, most closed system survive and reap the economic benefits inherent within that victory.

Aside from many stories and much pertinent information, I have retained one very useful possession of Jac Gadwill's—those exceedingly warm socks. It was those same thick wool L.L. Beans he'd loaned me, which I'd forgotten to give back, that I slipped onto my feet a few months after my return from Alaska. I then donned a pair of chest waders and stepped into the swift current of New York State's Salmon River. After so many months of researching salmon, watching other people catch salmon, and comparing how different types of farmed salmon stress the environment, I'd had enough. I wanted to get back to the reasons I became interested in fish in the first place. I wanted to catch a salmon.

Thirty years ago this would have been impossible in the Salmon River. Just as they were eliminated from Connecticut, salmon were eliminated from New York back in the 1800s. Many attempts to reintroduce them to Lake Ontario failed miserably. A lot of this was due to a profound shift in the environment. Industrial and agricultural runoff had fouled the water. The native freshwater herring runs that salmon had dined on had been displaced by alewives, a small seagoing fish that had invaded the Great Lakes with the opening of the St. Lawrence Seaway. With no predators to speak of, the alewive populations would soar and then die off in huge numbers when algal blooms caused a seasonal deoxygenation of the water. In the summertime along the shores of Lake Ontario the stench was horrific. A trip to the beach was a dreaded prospect for children all along the lake's coastline.

It is a different Lake Ontario and a different Salmon River now. With my pole in my left hand and my right grasping a tree branch for support, I pulled myself up out of the current and onto a rock, then paid out enough line for a cast. The fall foliage was in full swing, and the river was crowded with other fishermen in identical gear, methodically flipping their flies upstream and following them with their eyes as they completed their drifts. Periodically a flush of water released by the dam south of us sent a surge of discarded Styrofoam coffee cups swirling downstream. A rusty shopping cart overturned in the eddy next to me tottered in the current, with several old fishing flies and a length of monofilament line ensnared in its metal grillwork.

It seemed at first like one of those days that fishermen rue—when men far outnumber fish. All the activity, the flailing of line, the sloshing of boots, the tying and retying of different lures—all of it ritualistic hooey, designed more to impress other anglers than to draw the strike of a fish.

But as my eyes adjusted to the autumn light and the shapes beneath the surface of the water came clear, a vision presented itself that was, for me, heart-wrenching. The piece of algae that fluttered in the current next to the rock I stood upon recast itself as animal and not vegetable. It was in fact the frayed pectoral fin of a king salmon, a thirty-pounder, lazing in the current, not unlike that king salmon I'd seen twenty years earlier, just as the Oregon wild salmon were dying out for good. And at once the truth of the river came clear to me—I could see that next to this salmon was another, possibly its mate, and next to her was another and another. The river was paved with them. A hundred fish within reach of a cast.

Alongside all the extreme laboratory-based selection that has occurred with salmon, there is a kind of hybrid of natural-unnatural selection at work here in the Salmon River. The salmon at my feet in the lee of the current were Donaldson-strain kings, bred in a facility near Seattle, Washington, from a wide range of many different strains. Several of those strains are now extinct in their native Pacific Northwest environments. The Donaldson is therefore a kind of genetic message in a bottle, an amalgamation of genes, lost and found, combined in such a way as to make the Salmon River habitable by salmon again.

Around the world, while salmon geneticists try to make salmon more and more efficient and fit for a tank, there is starting to emerge a kind of reverse engineering in which wild-salmon advocates are applying more science-based methods to make tank-reared salmon fitter for return to the wild. In rivers where salmon had gone nearly extinct, like the river Tyne on the northeast coast of England, salmon rehabilitators are starting to find that the genetic complexity we have lost and fetishized over the last half century may not necessarily be the only key for staging a wild salmon resurrection.

Less than fifty years ago the Tyne was in the most dismal state of all United Kingdom salmon rivers. Its proximity to the industrial town of Newcastle, combined with a dam thrown across the river to create the Kielder Water reservoir, had destroyed the salmon population entirely. Not a single salmon returned to the Tyne in 1959. It might have stayed in this condition had it not been for a biologist and sportfisherman named Peter Gray, who decided to go against the popular conclusions in the arguments about salmon and genetics.

"If we go back to just after the last ice age," Gray wrote me, "all our salmon rivers *had* to recolonize. The genetic integrity had to start all over again." Salmon rivers were wiped out by glaciers throughout their range between ten thousand to twenty thousand years ago. Somehow, from a small genetic redoubt, they were able to reclaim their kingdom. There is a metagenetic component that must be respected, Gray agrees. West-coast Scottish salmon "turn right" to go north to Greenland, whereas east-coast "turn left." Putting a west-coaster in an east-coast river could send fish on a deadly holiday to France.

But if you have these metacomponents correct, you can start to goose salmon back to viability. Gray believes that we must get away from the mammoth hatcheries and industrial hatching facilities the salmon-farming industry helped concoct. Genetics are important, he agrees, but he has found that properly preparing juveniles for reintroduction and timing the stocking of rivers is even more so; it means the difference between success and failure. Hatchery-born salmon, it turns out, have to be taught what it's like to be wild again in order to make it. Gray introduces strong riverlike currents in their larval tanks. He feeds them insects and other food they will encounter in the wild when reintroduced to the river. And he releases them at a time when he knows other predators in the river will be largely absent or not feeding. All this has meant a complete reversal in the fate of the Tyne. Within thirty years of starting his efforts, he has brought the Tyne to the point where more than twenty thousand adult salmon return to spawn every year.

Salmon are inherently fragile, but also perhaps inherently resilient. Most salmon rivers were ruined at a time when we did not know how to mitigate our impact. But now we do. And if we can clean up rivers and make salmon-friendly conditions more possible in their former range, perhaps we will see wild salmon again in our lifetimes. In New York's Salmon River where I stood, I saw the evidence of this possibility with my own eyes. The Donaldson salmon that were stocked in the 1980s were originally put into Lake Ontario and the Salmon River for neither food nor sport, but rather to try to deal with the stinking mass of alewives that washed up on local beaches every summer.

The Donaldson fish did just that. But they grew big and powerful and beautiful, and fishermen wanted to catch them and eat them. The only problem was that Lake Ontario had suffered from nearly a century of industrial pollution, pollutants that ranged from persistant heavy metals like chromium to the manufacturing elements of the Vietnam-era defoliant Agent Orange. The fish were dangerously toxic. Fish and Game was ordered to stop stocking Donaldson kings into the Salmon River because of the health risk they posed to fishermen should they eat their catch. But even after Fish and Game stopped stocking them, something unusual happened. The Donaldson kings started spawning naturally. They had gone native.

The purist in me, the fisherman, the seeker of truly wild fish, wanted to recoil. What were these salmon at my feet? What would they become? What were Pacific salmon doing in a habitat that should be ruled by Atlantic-strain fish? What good were they to anybody if you couldn't eat them? All this went through my head until suddenly one of them, a magnificent golden brown animal four feet long and nearly a foot across the shoulders, reared up out of the water and grabbed my lure, pulling me off my rock with the force of its run. It was the pull of something wild. Something that dragged me upstream from my depressing thoughts of vanishing fish, suggesting that all was not over with salmon in my life. A mental adjustment would have to be made, but it seemed wrong to deny the presence of this salmon, an undeniably powerful and beautiful fish, in a river that twenty years earlier had been entirely devoid of them.

"To hell with it," I said to myself as line screamed off my reel and my heart beat and I chased the big fish up the river. The Salmon River, after all, should have salmon in it.

Sea Bass

The Holiday Fish Goes to Work

When you ask most seafood eaters which fish are farmed, most will say "salmon." Beyond that, consumer knowledge gets fuzzy. People seem to have a vague awareness that fish farming is growing, but why, where, by how much, and through what means remain under the radar. Consumers' default assumption still seems to be that a fish on the plate is most likely going to be wild. This in spite of the fact that aquaculture is the fastest-growing food-production system in the word and will likely surpass wild production within a year or two (if it hasn't done so already).

Take my stepmother, for example. About five years ago she told me she had found a new favorite fish. She'd eaten it on a recent trip to Italy, and she was happy to discover on her return that the fish had just become available in many upscale Italian restaurants throughout New York City. Lunching with her one day, I finally got a look at this new animal. It was called "branzino" on the menu, and, in the style of European seaside restaurants, it was to be served grilled and whole. Before it hit the flames, the waiter brought the fish out so that we could assess its freshness and quality and perhaps also to give us the impression of a holiday meal at the shore. My stepmother did not enjoy this part—her diet is, with the exception of fish, vegetarian, and she does not like to be confronted with evidence that fish are animals, with eyes as intelligent-looking as any mammal's.

The branzino was very fresh—its gaze was clear and its scales clamped down tightly against its flanks. It was exactly the size of a dinner plate, silver in color, with an attractive streamlined profile that reminded me of the American striped bass—perhaps the most famous game fish in America and a fish that has grown scarce on American menus since commercial fishing for it was temporarily banned in the mid-1980s and then severely restricted thereafter. A look through one of my fish atlases at home later that day revealed that the branzino was indeed very close to a striped bass—some taxonomists had even moved it over into the same genus as the striper—*Morone.*

The British call branzino "bass" or "sea bass" or "European sea bass," and they pursue them in their wild form as ardently as I had hunted striped bass in my youth. And once I'd made the acquaintance of European sea bass / sea bass / branzino, I found that I came across it everywhere. In the dozens of faux-French bistros that had sprouted up in America's urban centers, it was called "bar" or "loup de mer." In Italian trattorias that claimed a southern or Sicilian provenance, it went by the name "spigola," while the Spanish served it up with yellow rice as "robalo." And always when it appeared, it was

brought out whole and fresh, eyes clear and intelligent, exactly the
size of a dinner plate. A European seaside holiday in the time span
of a single meal.

Where had this fish come from? "Europe!" I was told by a variety of waiters. Where in Europe? No one quite seemed to know.
"Maybe the Mediterranean?" Why were we eating this European
bass instead of a local fish? "Because it's European!" seemed to be
the most common answer.

During the next few years, I was to become more intimate
with the European sea bass. I was to find that the taming of the European sea bass was one of the most important developments in the
relationship between humans and fish. For the forces that brought
the plate-size European sea bass to restaurants around the globe
represented the next phase in both managing and domesticating the
oceans. Unlike salmon, which adapt relatively easily to a farmed environment, sea bass and the wide range of ocean fish that we eat are
difficult to master. Their early lives are microscopic, their breeding
habits complex, and they seem inherently resistant to our designs of
putting them into our underwater mangers. That sea bass were
pulled out of the vast background of wild fish and eventually tamed
turns out to be the end result of a two-thousand-year-old process of
exploitation and scientific investigation, one that involved the efforts
of ancient Roman fishermen, modern Italian poachers, French and
Dutch nutritionists, a Greek marine biologist–turned–entrepreneur,
and an Israeli endocrinologist. All of them advertently and inadver-
tently created the conditions for an endgame that resulted in the
globalization of the European sea bass.

I was also to learn that by charting the history of European sea
bass and the word "bass" in general, I could get a sense of what had

happened to the world's coastal fisheries in the last quarter century and how human beings have gone about laying the groundwork for the next great artificial selection.

W hat do fishmongers and restaurateurs mean when they encourage us to choose something called bass? And why do so many fish seem to be lumped under that single name? The answer brings us back to the persistence of the primitive relationship between fish and fishermen and to the superstitious, highly unscientific way humans distinguish "good" edible fish from bad ones.

The English word "bass" derives from the Germanic *barse or barsch*, meaning "bristle" and most likely refers to the five-odd spiny rays that jut out from the dorsal side of species bearing that name. But as Anatoly Liberman, author of the book *Word Origins and How We Know Them* and one of the world's leading experts on names and their derivations, told me, fish names are slippery and not necessarily married to any one characteristic. "Several different fishes may have identical names, whereas similar-looking fishes may have widely different names." The name "moonfish," for example, applies to fish in at least seven different genera around the world. Many moonfish are roundish and vaguely moonlike, but many are not. As Liberman explained, this may stem from the fact that the very nature of hunting wild things compelled humans to be tricky and evasive with the names they chose. "Hunters and fishermen are superstitious people," Liberman continued, "and often prefer to call their potential prey in some indirect way, so that it won't hear and recognize the word."

Nevertheless, as humans emerged from prehistory into an age

of reason and classification, a somewhat primitive reliance on outward physical appearance rather than evolutionary provenance has
up until the twentieth century guided taxonomists. In the modern
era, everything commonly called bass, be it a European sea bass, an
American striped bass, or a Chilean sea bass, is classified as belong-
ing to a single scientific order, the order Perciformes, whose root,
perc, drives the researcher back to the Greek *perkē*. When that is
combined with the Latin *formes*, we end up with a classification that
means, broadly, "perch-shaped." Many fish turn out to be "perch-
shaped"—Perciformes is the largest order of vertebrates on earth,
containing over seven thousand species and most of the so-called
game fish of the world. It is so large a classification that taxonomists often call it a "garbage-bag holder," used to contain a ridicu-
lously large number of vaguely similar species that people haven't
quite gotten around to properly classifying. Curiously, and perhaps
not altogether coincidentally, the order Perciformes includes most
of the fish in the sea that people of European descent consider edible. "If it's perchlike," the classification seems to be saying, "let's
eat it."

Why we originally chose to eat so heartily from the order Perciformes is connected to evolutionary advancements that date back
250 million years. Whereas more primitive fish must constantly
swim to keep from sinking to the bottom, the forebears of the perciforms perfected an organ called the swim bladder, which they
inflate with gas to keep them neutrally buoyant in the water column,
much as a scuba diver inflates a buoyancy compensator to achieve a
state of weightlessness. When a perciform dives deeper, it emits
more gas into its bladder, which compensates for the added pressure
of the water above. When it rises, it absorbs gas back into its tissues,

once again finding a weightless equilibrium. And, like a scuba diver who has properly adjusted his buoyancy compensator, a fish that has achieved neutral buoyancy expends less energy.

The perciforms' victory over gravity has in turn led to other morphological adaptations that make them both successful animals and good to eat. Without a need to fight gravity all the time, perciforms became more efficient swimmers and were able to trade in their heavy, energy-demanding "red muscle" tissue for lighter, more delicate flesh. Hence the white, light meat of many perciforms. Perciforms also evolved an efficient muscle structure that is principally attached only to the central spinal column. The result: a smooth, mostly boneless fillet, very pleasant to eat.

The last way the perciforms' swim bladder makes them attractive to us as food is not the possibilities it gives them but rather the *limitations* it imposes. Going back to the scuba-diving analogy, there is only so deep a diver can go before his buoyancy compensator becomes useless. Below this depth, water pressure will overwhelm the gas inside the compensator and the device will implode, making the diver sink like a stone. Fish equipped with swim bladders have the same problem and are therefore limited to a certain depth range. Is it a coincidence that the maximum depth to which coastal perciforms can venture is similar to the depth to which a human free diver can swim or an early human's primitive fishing line can reach? It could be that the fish we have come to recognize most widely as being edible are the ones that primitive Europeans could most easily catch.

But to get back to why one perciform in particular, the European sea bass, ended up as my stepmother's fish of choice, other issues need to be considered—most important, issues of fish scarcity and human abundance.

As early humans perfected their ability to catch perciforms, they started to affect those fishes' numbers. And the perciforms that disappeared early on became more prized. Evidence of this is the way humans began to endow scarcer fish with positive anthropomorphic characteristics, especially in places where the fish was subject to more intense fishing pressure. In the Anglo-Germanic part of their range, where human populations were sparse and fishing pressure modest, sea bass were called "bass"—that is, basically "prickly." But in Mediterranean Europe the same species began to be named in a way that indicated agency and intelligence. The ancient Greeks associated the fish with the word *labros,* or "turbulence." Homer uses *labros* in reference to wind and water, and later authors use it about people, in the sense of violence or boisterousness. But *labros* as it applied to sea bass gradually came to imply cleverness. In modern Greek the concept of the sea bass as a clever fish became its defining characteristic. Today the fish is called *lavraki*—"the clever one." If you wanted to indicate in modern Greek that someone had cleverly figured out something tricky and challenging, you would say that he *epyase lavraki*—"he caught a sea bass."

The perception of the bass as clever occurs in other Mediterranean languages. The Romans named the fish after an animal they considered particularly intelligent—*lupinus,* which eventually became the French *loup de mer*—"sea wolf." And the Latin poet Ovid wrote of sea bass as using its smarts to frustrate its potential captors. "In vain above the greedy [fisherman] toils," Ovid wrote, "while with arts more exquisite the bass beguiles."

European sea bass thus seem to have rapidly solidified their reputation for cleverness in the Mediterranean. The reason for this may be a direct product of the holiday-like environment of the Mediterranean Sea, the place where humans and sea bass had their most intense interactions. The Mediterranean occupies an exceptionally warm and dry climatic zone. Most rivers on the European continent flow away from it, meaning that, compared to other seas, the Mediterranean's biotic systems receive few nutrients. The sea is therefore described by scientists with the Greek-derived word *oligotrophic*—a place that "contains little nourishment." This oligotrophia begins at the bottom of the food chain: only a sparse amount of phytoplankton is able to exist on the thin supply of wastes being washed in from land to sea. And with a very low food base at the bottom of the ecological chain, as you go up the chain each level is thinner than in most coastal areas. By the time one reaches the level of the European sea bass, both the population of fish and the size of individual fish are naturally smaller and more sensitive to overexploitation than in more productive seas. Recently the Italian cookbook author Marcella Hazan told me that when she moved to the United States, she simply could not find the right fish for her European sea bass recipes. "Your bass are too big!" she lamented.

The Mediterranean is also exceptionally deep—over three miles at its abyssal point near the Bay of Pylos. This makes it capable of retaining heat and keeping coasts warm even in times of extreme cold. During the last ice age, Mediterranean peoples not only survived but probably increased their numbers while the rest of Europe's population dwindled in their caves.

The presence of a consistently expanding human population combined with a sea that is inherently stingy as a food source created the conditions for an unbalanced man/fish relationship dynamic, a dynamic that would get continually more unbalanced as humans became more numerous and better at fishing. And it would finally reach a point where the man/fish imbalance was so great that something other than fishing had to come along to correct the disparity. The Mediterranean is many years ahead of the rest of the world's oceans in terms of the rise of man and the fall of fish. The decline of wild fish stocks in the Mediterranean in the last half century and the subsequent attempt to repopulate the sea through aquaculture is a stark forewarning of what could happen all over the world as the oceans in general become oligotrophic—containing little nourishment for everybody.

How did early humans choose the animals they were going to tame and eat? An examination of middens at Neolithic European dwellings reveals that humans used to eat from a fairly wide buffet of wild game. In varying amounts, the meats they consumed consisted of red deer, boar, cow, roe deer, horse, goat, antelope, elk, chamois, bison, reindeer, fox, badger, cat, marten, bear, wolf, dog, otter, lynx, weasel, mouse, rat, rabbit, beaver, and marmot. By the time of Christ, we were down to four basic kinds of mammals in our fire pits: sheep, goats, pigs, and cattle.

When it came to birds, there was a similarly broad choice available. Pigeon, snipe, woodcock, pheasant, grouse, dozens of different ducks, grebe, various wading birds, and many more. Today we focus on four primary birds for our food: chickens, turkeys, ducks, and geese.

Why those animals?

Irrespective of his infamous eugenics writings, anthropologists consider the list of criteria set out by the nineteenth-century intellectual Francis Galton as being a good thumbnail sketch for what guided Neolithic humans. Of the animals that humans chose to domesticate, Galton believed that they must have been:

1. hardy
2. endowed with an inborn liking for man
3. comfort-loving
4. able to breed freely
5. needful of only a minimal amount of tending

When wild fish shortages in the Mediterranean began to be apparent in the 1960s, this list of criteria was readily available to anyone willing to do a little library research. Yet in the end the list was pretty much disregarded. The mid-1960s were the peak of the so-called Green Revolution, a time when great faith was put into often radical scientific techniques for boosting food production. This techno-enthusiasm held much greater sway over the early ocean-farming researchers than did the ancient choosing habits of Neolithic man. As an American ecologist and aquaculturist named Josh Goldman noted recently, "Fish farming has the misfortune of coming into being at a time when all sorts of science was at its disposal." Through the application of modern research, scientists believed that any species could and should be tamed, regardless of how out of sync that species' traits might have been with basic principles of domestication.

Witness the European sea bass. If researchers had thought to measure the fish against Galton's criteria, there is little doubt they would have chosen differently. It is a failure in every category:

1. **They should be hardy.** European sea bass are not. Sea bass, like most ocean perciforms, lay over a million eggs, but out of this plentitude only one or two become viable adults. Not only are newly hatched fish extremely fragile, but there is a whole phalanx of diseases that attack them, particularly at larval stages of development.

It is very difficult to make a long-term investment in a species' welfare if the natural state of things is to have more than 99 percent of a population die.

2. **They should have an inborn liking for man.** Sea bass and most marine perciforms are at best indifferent to us. While there are early examples preserved in the fossil record of Neolithic man's having tied ropes around the jaws of bears and early horses and leading them around as pets, there is nothing quite analogous with marine fish. No early human ever went down to the sea and brought back a pet sea bass. There is, of course, a long tradition of freshwater aquariums and ornamental fish. But aquariums are simply moving landscapes—one-way interactions where little is expected of the domesticated animal other than color and motion.

3. **They should be comfort-loving.** Sea bass and marine perciforms generally are not. They are responsive to easily available food, but many hate containment. Early attempts at taming desirable fish like the Mediterranean dentex produced dismal results—the dentex sulks listlessly at the bottom of a net pen and refuses to eat. Other perciforms, like the mahimahi, or dolphinfish, slam themselves repeatedly against their containment buoys or shred themselves to death against the netting of their pens.

4. **They should breed freely.** Understanding mammalian reproductive systems was, even for primitive man, a straightforward thing to do. Coitus could be witnessed, as could the birthing of juveniles. But sea bass reproduction is mysterious, taking place largely outside the bodies of the fish with sperm and eggs that are nearly microscopic. Of course, fish reproduction is, as it is with all vertebrates, a matter of merging sperm and egg, but even with that understood, merely putting a male and female of the same species in a tank

does not induce breeding. Most marine perciforms shut down their reproductive activity completely when brought into a captive environment.

5. **They should be easy to tend.** Sea bass, when they are born, are distinctly unprepared for life. Domestic mammals and birds pass through their larval forms inside the nurturing body of a mother animal or within the hard, self-contained world of a nutrient-rich, calcium-based egg. In those cases humans do not have to meddle with the subtler, microscopic developments that transform a small cluster of cells into a complex, fully formed, and even self-reliant organism. Even most freshwater fish are better than sea bass; trout, salmon, catfish, and carp all hatch out of large, nourishing eggs, and after the young are born, the larvae emerge with a significant egg sac attached to their abdomens. Contained within that yolk sac is enough nutrition to last the first few weeks of free-moving life. Ocean perciforms, meanwhile, are completely defenseless. Since the game of marine fish is to play long odds, parents invest little into each individual egg. What scant nutrition exists is a bit like the few gulps of oxygen a passenger on a falling airplane might get in an emergency landing—enough to survive a transition but nothing to rely on beyond that. And so sea bass must find prey from the very early days of their existence. When the fish are in captivity, humans must re-create a whole parallel microscopic rangeland for these tiny hunters to make their first kills.

In fact, if you were to look for a portrait of an animal that by all rights *shouldn't* be domesticated, you would be hard-pressed to find a better example than the European sea bass for your case study. The logical decision would have been to seek out and tame a more naturally suitable animal. But Europeans already knew and liked the sea bass, perhaps more than any other European coastal fish. Even though it would take a substantial investment of money and intellectual energy to "close the life cycle" (as aquaculture scientists call the process of complete domestication), early aquaculture pioneers decided the sea bass was the best place to start.

Like the airplane, the telephone, the incandescent lightbulb, and all other great technological leaps of modern history, the domestication of the sea bass specifically and ocean perciforms in general has a host of nations and individuals who claim to be the responsible party. Postwar Japan in the 1940s launched an effort to tame a famous but declining perciform—*Pagrus major*, or red porgy, another "holiday fish" that any self-respecting Japanese is all but required to have at a formal wedding banquet. In Europe, France worked intensively in the middle years of sea bass development in the 1970s, alongside serious study of a flounderlike fish called the turbot. But if necessity is the mother of invention, then an Israeli mother can lay a pretty heavy claim to birthing the domesticated sea bass.

Long before other Mediterranean people felt that their wild fish had gone into enough of a deficit to consider replacing them with farmed sources, Israeli leaders were deeply aware of their nation's shortages. "Food security" and "food sovereignty," two terms that are today frequently discussed by agronomists and ecologists in the rest of the world, were the fundamental terms of survival decades earlier in Israel. Beginning in the 1950s, Israelis developed a multipronged approach for the development of homeland-produced food that involved a network of government research centers and kibbutz farm complexes.

This applied approach to food development research was particularly successful with fish farming. The tight relationship between government institutions and working kibbutz farms meant that experimental projects could be instantly field-tested. This applied research had immediate payoffs. In the 1940s, collecting know-how from the postwar European diaspora, Israelis began cultivating carp, building on a four-thousand-year tradition that had been borrowed from China. Carp are not endemic to Europe. Their cultivation began with the opening of trade with China in the late Middle Ages, which had been launched in Europe to address a fisheries crash that had preceded the declines in the Mediterranean. Europeans had overfished their freshwater lakes and rivers since Roman times and had gradually turned some freshwater ponds into carp farms. But the European carp culture had always been something of a cottage industry; carp flesh had many small bones dispersed throughout and was often used by poorer folk in shtetls who ground it up into gefilte fish and other forms of processed fish meat. By 1948, in sharp contrast with the rest of the world, which continued to rely on wild fish, more than 70 percent of the fish that Israelis ate were farmed. Most of that farmed fish was carp.

But carp live in fresh water—an extremely scarce commodity in Israel, often the very casus belli of regional conflict. What Israel did have was an abundance of salt water abutting its long seacoast. For a country with a total land area of only eight thousand square miles, an additional strip of potentially bountiful ocean territory *had* to be used.

In 1967 another large resource fell into Israeli hands that would give them one more advantage in launching an ocean farming project. After a period of rising tensions that began, ironically, with a dispute over rights to fresh water in the Jordan River, war broke out between Israel and its Arab neighbors. In what became known as the Six-Day War, Israel fought off an Arab offensive from all sides and in a counteroffensive wrested control of the Sinai Peninsula from Egypt. After the capture of the Sinai, religious Jews rejoiced that Israel now controlled the mountain where God supposedly gave Moses the Ten Commandments. Fish researchers (generally some of the more secular people you are likely to meet) had no interest in the mountain. They made for the sea and the rich waters of the Sinai's coastal plain.

On the northern edge of the Sinai lies Lake Bardawil, a sea within a sea formed by a spit of land that stretches semiporously across the top of the peninsula. This extremely hot, salty water is a holy land for European sea bass. Because it has little exchange with the greater Mediterranean and is located in a region of intense sunlight and rapid evaporation, Lake Bardawil is a zone of hypersalinity—a condition critical to spawning sea bass. It was on Lake Bardawil that a young endocrinologist named Yonathan Zohar was to begin research that would shape the fate of sea bass and, in fact, all fish for many years to come.

If one were given over to poetic comparisons, Zohar might be likened to the Greek god Eros—the god of love and fertility. In reality Zohar is very concrete and empirical and uses a more scientific analogy to convey exactly what it is that he does. Recently he figured out a simile that works particularly well for his ninety-year-old mother. "I tell her," Zohar explained, "that I am like ob-gyn for fish." Over the years he has gained a reputation among aquaculturists as one of the world's best at cracking the reproductive codes of the marine world.

Soon after the Israeli annexation of the Sinai, Zohar began his graduate studies at the National Center for Mariculture of Hebrew University. Situated at the very extreme southern end of present-day Israel's territory in the city of Eilat, the center was charged with developing marine aquaculture focusing on the Red Sea. It was the very Green Revolution sentiment of alleviating hunger that motivated Zohar in his choice of disciplines. "I wanted to do something," Zohar told me, "that would help feed people."

But why did the Israelis start with sea bass? The answer appears to be its potential not only as a food source but also as a commodity. A French fisheries policy analyst told me recently, "Traditionally, sea bass was rare. It was a big fish that you had for a special party with friends and family." This was a tradition that dated back to ancient times. Romans believed that the fish should be cooked whole, ungutted, and the head was particularly prized.

Romans probably made their case for sea bass throughout the continent, for European sea bass have an exceptionally broad range— inhabiting the entirety of the Mediterranean on out to the Strait of Gibraltar and up the coast of Spain, France, England, the Netherlands, and the Baltic coast of Germany. In a way it is a kind of fish version of the euro, a valuable silvery commodity that finds its way into near-shore pockets of virtually every continental country's coast. It had in effect marketed itself to Europeans for centuries. Israelis realized that developing a marine fish as a viable domesticated product would take a huge investment. At least some prospect of selling the fish at a profit had to be on the horizon. With such a large international reputation and the added value of being considered a holiday fish for which people would pay high prices, the sea bass seemed, in spite of its biological limitations, the best one to try.

But in order to make this holiday fish into an everyday fish, several difficult constraints had to be mastered. Since no one had figured out how to make sea bass spawn in captivity, Zohar and his colleagues had to collect fish from nature for their research. This is where Israel's acquisition of Lake Bardawil became of critical importance. Early on, Jewish researchers passing through the newly captured Arab land realized that Lake Bardawil was a gold mine for research.

"It was very difficult," Zohar recalled. "You used to go to the Bardawil, travel for days and collect the young fish, and you'd truck them back across the desert all the way to your tanks in Eilat and start to grow them, which, if you think about it, is a pretty crazy way to do aquaculture." Many uncomfortable days in those trucks for Zohar, a tall, lanky man with a chronically bad back, eventually sparked some serious rethinking. Collecting fish from hostile Arab territory and then growing them in Israel was ridiculous.

For a while Zohar and his colleagues tried replicating the environment of Lake Bardawil. But nothing worked. And so they looked at the problem from the biochemical side. "We developed a hypothesis," Zohar recalled, "that the lack of the correct spawning conditions is transduced into a *hormonal* failure, . . . and so we decided we had to go into that hormonal system and try to see what was going on."

Getting to the hormonal center of a sea bass and the other perciforms being investigated at Eilat was not an easy thing. Sexual hormones are manufactured in a fish's pituitary gland, an organ about a quarter of the size of a pea, located inside the bony casement of the fish's skull. Over the course of the next ten years, Zohar led a team that picked through tens of thousands of fish heads, tweezing out the tiny pituitaries and sending samples around the world for analysis.

'There is a natural human tendency to want to create linear narratives out of the disorder of day-to-day life. But science is far from linear. The understanding of perciform reproductive systems took a decade of false starts. At one point, after a particularly arduous harvesting of ten thousand pituitary glands, a laboratory where Zohar had sent the samples for analysis called to tell him that all the pituitary material was "degraded." Years later Zohar still raises his bushy white eyebrows recalling the frustration. "I tell her, 'No, no, no, no, no. It can't be! I collected it as fresh as it can be!'" Zohar spent another twelve months harvesting another ten thousand pituitaries, only to get the same answer from the lab.

Zohar's career itself might have "degraded" into an endless spiral of worthless research at that point, had he not decided to revisit the results. When he did, he realized he had unlocked not just the secret of sea bass reproduction but a basic problem that underlay all questions of fertility, even human. Zohar realized that the spike noted in the lab's analysis showed a profound chemical shift. The sample wasn't degraded; an additional hormone—a new one that no one ever before knew existed—was present in the fish's pituitary during spawning, which had thrown off the results.

The problem seemed solved. But again, perciforms are tricky. Within a few hours of injecting the fish with the spawning-inducing hormone, the chemical would completely disappear from the fish's bloodstream. It seemed that a "cleaving enzyme" was the problem; there appeared to be a compound manufactured inside the wild fish, always at the ready to dispel the effects of the hormone. Eventually Zohar realized he would have to *manufacture* a whole new hormone that was impervious to the enzyme.

Making the hormone took even more analysis and years of false starts. But when it was achieved, not even this was enough. Sea bass are "asynchronous spawners," fish that hold on to their eggs for many days so that they may spread them out over a variety of terrains and conditions. It was therefore necessary to hormonally convince sea bass that *now* is the time, *this* is the place—lay *all* your eggs at this very moment, because this is the best bet you're ever going to get in your entire life. Eventually Zohar's lab engineered a microscopic polymer-based sphere that would slowly release the hormone into the sea bass's bloodstream, causing the fish to expel all its eggs and sperm in a single, predictable period.

By the 1980s, sea bass fertility had finally been decoded. But by the time all this had been achieved, times and world events had shifted and the window that would have allowed Israel to corner the sea bass market began to close. The Sinai Peninsula was returned to Egypt in 1979, and with it Lake Bardawil and all the potential research it represented. While scientists like Zohar had been launched in their careers and their scientific discoveries by a bold, food-desperate Israel, that nation had lost its impetus by the 1980s. The survivalist kibbutzim of the early idealistic days were fading. Scientists like Zohar left for positions abroad.

The task of launching the sea bass on its global odyssey, an odyssey that would lead to my stepmother's plate in New York, would have to fall to another country. As luck would have it, it fell in 1982 to the nation where odysseys were invented.

If Yonathan Zohar is the Eros of sea bass, then it's fair to consider a tall Greek man named Thanasis Frentzos as the fish's Odysseus.

Thanasis lives on the island of Cephalonia, located midway up the western coast of Greece. Within sight of Odysseus's home island of Ithaca, Cephalonia is emerald and olive in its highlands, with wide, foaming rivers that empty into the bluer-than-blue Ionian Sea. In recent years archaeologists have put forth the idea that Cephalonia may in fact have been the real home of Odysseus, that the small island next door is an impostor island—a lesser remnant from a volcanic explosion that broke the original, much larger proto-Cephalonia into a half dozen pieces. When flying over the island it is easy to imagine the place as being the "sea-girt land" of Odysseus's kingdom, one Homer aptly described as "a rugged territory, and yet a kindly nurse."

Thanasis is similarly Homeric. Trim and square-shouldered, with a deep, resonant voice, a flowing mane of hair, a heroically long and flat nose, and an impressive, curling beard, he seems like someone recently escaped from the side of an ancient urn, sprung into life with all of Odysseus's wily enthusiasm. But where Odysseus set sail on an eastward course from his home island toward Anatolian Troy to reclaim a beautiful woman, Thanasis in 1982 headed west, to Sicily, to bring back twenty thousand European sea bass.

Just as Helen was snatched away from Greek shores by foreigners backed by the power of the gods, the few remaining pockets of wild sea bass in western Greece were, according to Frentzos, stolen throughout the 1960s and '70s from Greek waters by foreigners backed by the power of dynamite. As the home sea of twenty-three different nations, the Mediterranean has since 1949 been regulated by the General Fisheries Commission for the Mediterranean (GFCM), one of the oldest regional fishing agreements in the world. The GFCM historically manages large migrating stocks of fish, like hake, that cross multiple borders. Member nations have been relatively compliant in jointly managing these "straddling stocks" of fish, and managers report that these populations of fish are in decent shape. But sea bass are considered "local" fish and are not overseen by the GFCM. As a result it falls to individual nations to protect stocks of sea bass. In the 1970s, while Thanasis was getting his doctorate in marine biology in New Zealand this oversight had major repercussions for Greek sea bass. For just as in ancient times, Greeks proved largely ineffective in defending their property from the wiles of Romans.

There is a general rule when it comes to overfishing. If no regulations are put in place, the more fish populations decline, and the more extreme and ecologically damaging fishing methods get. Extreme methods are considered necessary to make a day's fishing worthwhile. Beginning in the 1970s, poachers speedboating over from Italy armed with dynamite sealed the fate of the bass in the Ionian Sea. Explosives tossed from the boats and detonated underwater would create violent pressure waves that overloaded the sea bass's neurotransmitters, knocking them unconscious and causing them to float out of the crannies of their home reefs. Unconscious, the bass were unable to regulate their swim bladders, and so their bellies would swell with gas and the fish would rise to the surface, where they were scooped up and taken back to Ancona and other towns along the Italian coast.

When Thanasis returned from his studies in New Zealand, he saw a very changed sea. Even during the prime sea bass spawning season, the so-called halcyon days of the January thaw, when the kingfisher birds would return to the Ionian Sea to lay their eggs ("halcyon" coming from the Greek word for kingfisher), he noticed very few spawning sea bass. When a sea bass did appear in the marketplace, it was an event. A whole two-pound fish could cost the equivalent of fifty U.S. dollars.

So Thanasis Frentzos decided to head to Sicily to find his own sea bass. For in the interim between when he left for New Zealand and when he returned, French scientists had picked up where the Israelis had left off and advanced sea bass domestication considerably. Italians were now employing some of the French technology in makeshift hatcheries, and Thanasis figured that Italy was the easiest and cheapest location from which to get the fish to Greece. Once he got them home, he planned to raise these fish to adulthood and begin a self-perpetuating colony of domestic sea bass on Greek territory. Perhaps he saw this as a way of restoring a piece of Greece's heritage, much the same as Greek nationalists at the time were pressing England to return the Elgin Marbles that Lord Elgin took from the façade of the Parthenon.

But Thanasis's goal was not only patriotic. The idea of a fisheries crisis was starting to move from being a uniquely Mediterranean phenomenon to being a global one. Anyone studying marine biology at that time could sense a change in the air. In 1977, in response to many nations' complaints that their fish were being "stolen," the United Nations passed a revised Law of the Sea Treaty that allowed countries to expand their sovereignty from three miles up to two hundred miles out to sea—the zone that most fish called "bass" were likely to inhabit. The United States aggressively developed its own treaty, called the Magnuson-Stevens Act, which effectively shut European nations out of New England's Georges Bank and other fertile fishing grounds up and down the Atlantic coast of the United States.

Conservationists were also starting to gain ground. In 1982, the same year that the International Whaling Commission passed a global moratorium on the hunting of whales, the eastern United States' favorite bass, the American striped bass, reached its lowest population levels on record, resulting in a similar moratorium several years later. American striped bass, a fish that early English settlers probably named after the European sea bass, had declined precipitously throughout the 1970s to their lowest levels in history. Thanks to a protracted effort led by sportsmen and conservationists, all fishing, sport or commercial, was banned for three years. Also in 1982, another popular fish called bass, the California white sea bass, was facing similar regulatory reductions. In that year the Mexican government would ban the United States from fishing for white sea bass in Mexican waters, thereby removing yet another bass from marketers' rosters.

In response, in 1983, still another "bass" was to appear in markets in Asia and America—a fish called the Patagonian toothfish, that sold poorly until it was renamed "Chilean sea bass." An international niche for a white, meaty, basslike fish was starting to open up around the world. And for Thanasis Frentzos it seemed that the chance to "catch a sea bass" in the literal and financial sense of the Greek expression was close at hand.

Greece in 1982 was not an easy place to finance a risky venture. A military junta known as "The Colonels" had been ousted less than a decade earlier, and the country was still seen as something of a European banana republic. But Cephalonia is known as an island of eccentrics and risk takers. Its patron saint is St. Gerasimos—the protector of holy fools. Thanasis was lucky enough to have a friend in Marinos Yeroulanos, a civil engineer–turned–entrepreneur and fellow Cephalonian. Yeroulanos loaned Thanasis the equivalent of two yearly salaries for a marine biologist, as well as his yacht, and, in the tradition of earlier Cephalonians, Thanasis set out on a thirty-foot sailboat with a tiny engine across a dangerous sea.

On arriving in Sicily, Frentzos stopped at the holding station where he was to pick up his fish. He was deeply unimpressed. The owners had gotten their hands on a load of three-inch sea bass from an Italian research institute and had reared them in a lagoon to the size where they could be sold. Loading up Frentzos's tanks, the Italians raised their eyebrows at his jerry-rigged boat but gladly accepted the payment—twenty-eight thousand dollars—a fortune in Europe at a time when banks were charging an average of 20 percent interest for business loans.

Sailing out of the harbor, Frentzos began to feel more hopeful. He calculated how much the load might be worth if he were to manage to get the sea bass home and grow them to their full market weight. After growing his twenty thousand little sea bass to market weight, about two pounds each, multiplied by fifty dollars per fish. . . . Could it really be a *million dollars*? Of course there would be expenses. And inevitably there would be some mortality.

But most fish mortality happens during the very first days of a fish's existence, a phase that Thanasis, a student of ichthyoplankton, knew well—the microscopic yet monstrous-looking forms fish take when they are first hatched. During his time in New Zealand, Thanasis had spent countless hours looking at these strange pre-fish fish through a microscope, watching them float across a two-dimensional plane that seemed a mile long, fluttering and spinning, breathing through their larval gills, dying when the fragile balance of food, salt, water, and oxygen shifted ever so slightly. All this examination was so new in the 1970s that he and his colleagues were constantly coming across previously unknown species. Thanasis, being the only speaker of Greek in the lab, was often called over to a colleague's microscope and asked to come up with a scientific name that matched the weird organism on the slide. "That dorsal fin looks like a bridge," one of them would say. "Thanasis, what's the Greek word for 'bridge'?"

Years of watching the dance of larval fish had taught him that nature's winnowing is most likely to take place at this delicate, hypersensitive stage. But later, off the coast of Sicily, his boat full of tiny sea bass, Thanasis felt he was past all that. The fish in his barrels lashed to the gunnels of his borrowed boat, were three inches long—they were *already* survivors. The Italians had already paid the price of the initial attrition in growing them out to fingerling size. In his view the Italians had taken the risk, and Thanasis was going to get the windfall by growing them out to market size and selling them at a tremendous profit. Poetic justice for a Greek who had seen his home sea emptied of wild bass by Greece's greatest classical rival.

All these thoughts must have been on Thanasis's mind as the mountains of Sicily faded into the distance. Which is when the wind started to blow. . . .

B efore a strong wind blows on the Mediterranean, a crystal-clear sky is usually observed over the dark purple blue plain of the sea. Such was the case as Sicily dropped behind Thanasis's boat. By nightfall all that had changed when a freak early summer storm hit. Soon a force-seven wind was blowing up behind the boat, driving the prow down under the waves. The sea bass in their ad hoc barrels were secured to the side of the boat with heavy chain and fine-meshed screens over the tops to keep them from sloshing into the sea. But the oxygen tanks that were aerating the water were listing in their housings and pulling at the hoses that led down into the water. Thanasis and the captain stayed inside the wheelhouse while the tanks clanked ominously against one another. If they were to become detached from their hoses, the fish would surely die.

Staring out the window of the wheelhouse at twenty-eight thousand dollars' worth of sea bass fingerlings proved too much for Thanasis. Holding on to the stays of the mainsail, he pulled himself out onto the deck of the boat. Waves running across the bow blew up into clouds of spray, sending stinging salt into his eyes. Gazing out over the chaotic scene, he could see what looked like an oxygen tank rolling back and forth over the hose that fed into the live wells. He crawled on his hands and knees and reached out for the tank. It rolled away from his fingers, then back, then away again. Finally he made one last lunge and held on to its edge. But just as he did so, the captain grabbed his shoulder and pulled him inside. "It's not worth it," the captain said. "It's not worth going to the bottom of the sea for a bunch of fish."

On and on the wind blew throughout the night, turning the boat round and round. All headings were lost in the wind, and the darkness and the night seemed as if they would never end. But end they did. A rosy-fingered dawn began to glow in the east. Thanasis and the captain staggered on deck.

"What is the first question a sailor asks when he is in trouble?" Thanasis wondered aloud to me two decades later, before deftly filleting a cooked sea bass we were sharing for lunch. He raised his substantial eyebrows for the punch line. "You might think this sailor would ask, 'Do I have anything to drink?' or 'Do I have any food?' But no. That's not what he asks. What he asks is, 'Where am I?'"

In the distance on that frightening June morning in 1982, they could see a headland of sorts, but it was unclear if it was an island or a continent. And since orientation has a strange persistence in the human mind (i.e., what started out as south on the port side must *still* be south, even though you have turned in a hundred circles since that initial orientation was fixed), Thanasis felt a knot in the his gut. He felt sure that this land could be nothing other than Mu'ammar al-Gadhafi's Libya, directly to the south of Sicily and a famed pirate haven at the time. How would he ever get the boat back to the Greek entrepreneur who had loaned it to him? Why the hell had he abandoned the life of a simple oceanographic researcher to become a fish farmer?

But as they approached the land, the outline of a familiar house came into view. It was a large, many-windowed cottage with the characteristically jade green Ionian shutters, still closed up tight from the previous night's storm. It looked just like a house Thanasis knew well, the house of an acquaintance named Claudatos, who lived on a promontory overlooking his home harbor. "My God," Thanasis whispered, "could that actually be Claudatos's house?" He waved and shouted, and soon Claudatos himself emerged onto the portico, waving back, the morning sun glinting pleasantly off his bald head. The wind, like some kind of Athena-driven lackey god, had blown the boat safely back to Cephalonia. Thanasis embraced his captain and breathed a vast sigh of relief. It was only then that he noticed that one of the oxygen tanks had come to rest on the aeration hose leading into the sea bass tanks. The hose was kinked, and the tank had pinched it closed. Peering apprehensively into the barrels, Thanasis saw, in miniature, the scene he'd beheld after the Italians had dynamited his home reefs: thousands of tiny sea bass, their swim bladders filled with gas, floating belly-up in the water, suffocated to death.

But Thanasis discovered that amid all these dead fish a few in each tank had survived even without oxygen. He counted the fish one by one. There were exactly 2,153 survivors. These few fish, selected by their ability to withstand stress and oxygen deprivation, were to be the founders of a global race of sea bass.

The ring tone on Thanasis Frentzos's mobile phone is an excerpt from a radio broadcast of the 2004 European Football Championship. In that game Greece, an eighty-to-one long shot, played the vastly superior Portuguese team to a 0–0 draw for almost the entire match. Finally, with just a few minutes to go, Angelos Charisteas sprang into the air and headed a corner kick into the right side of the net. "Goaaaaaaallllll! Greece one, Portugal zero!" screams Thanasis Frentzos's mobile phone when someone wants to get hold of him.

A last-minute victory against long odds is exciting to all small nations. But Greeks feel such a victory with particular pride and sense of justice. As the very *founders* of the scientific method, many Greeks believe they *should* lead the world, and after so many years as Europe's lowliest economy, they rejoice heartily at any successes that come their way.

Such was the victory Thanasis had been hoping to pull off on his trip to Sicily. Though much of the work on sea bass taming had been done prior to his trip, the industry had yet to take off, and here, Thanasis thought, was his opening. He set about raising his 2,153 fish in a manner that would maximize their survival.

Early on, though, he encountered a problem he couldn't seem to solve. After the first breeding of the initial tribe of fish, nearly 50 percent emerged with crooked backbones. While the fish were perfectly edible, they were unappealing, especially in a culture that prefers to have its fish served whole. In fact, the weird shape of these first cultured marine fish was to give rise to fallacious speculation throughout Europe that fish were being genetically engineered. It was a serious problem, and Frentzos applied his small staff to solving it as quickly as possible. "We couldn't figure out why it was happening," Frentzos recalled. "Is it cancer?" he wondered. It got to the point where desperate ideas were being thrown out. "I know," Thanasis ventured at one point, "the fish don't have enough vitamin C!"

Indeed, besides breeding, nutrition had been the next biggest bottleneck in taming sea bass. When they emerge from their eggs, the lack of a significant yolk sac makes them extremely vulnerable. They must immediately find something to hunt. But because they are so underdeveloped, lacking functioning eyes and equipped only with rudimentary nostrils, the only way they can locate prey is by using their neuron-rich lateral lines to sense the vibrations prey creates when it moves. In nature, sea bass are born right after the hatch of phytoplankton, microscopic algae that in turn act as fodder for tiny animals called zooplankton. The zooplankton wriggle with vigor during the halcyon days, luring sea bass to hunt them down.

The logical thing for fish farmers to have done would have been to grow zooplankton in captivity. But in some ways, zooplankton is as difficult to domesticate as sea bass. Eventually a different creature was found that could in effect merge the phytoplankton and zooplankton food chain into a single link. This class of creature was the freshwater animal called the rotifer. Initially considered a nuisance species that plagued Chinese carp growers, fouling their ponds and circulation systems, rotifers were at first merely skimmed off the surface of the water and discarded. But eventually early fish farmers in Japan (another nation with extreme food-security concerns) realized that they could be used to feed very small juvenile fish.

It was in France and Holland that the rotifer was perfected as early sea bass food. Pascal Divanach is a merry Frenchman hailing from Brittany who now makes his home on the island of Crete in Greece. Divanach descends from an old family deeply connected to the seafood industry. His brother has grown wealthy feeding fish to Europeans through the Clemon Accord Group, a seafood restaurant chain that is renowned throughout France. But it was Pascal and his fellow aquaculture researchers in France and Holland who figured out how to feed the fish that now feeds many Europeans.

Divanach was invited to the Greek Institute of Oceanography in the 1980s and married a Greek woman soon after. Still proudly French, he delights in his adopted country and has a personal sense of pride in the Greek fish-farming industry. While seated in his office outside the town of Iráklion, he held up a promotional sticker for Greek aquaculture that said, FISH OF GREECE, FISH OF THE SUN. "It's very beautiful, don't you think?" He went on to explain to me what the French contributed to the taming of the sea bass.

"The big advancement for sea bass culture was something we call the green-water effect," Divanach told me. "In early systems they would introduce the rotifers and let them bloom. Because it was a closed system, bacteria would accumulate after twenty days and spread to the juvenile sea bass. In France we opened the system slightly. For one month the system was open to the sea to allow the refreshment of the environment with phytoplankton. This led to more food for the rotifers and better health and nutrition."

It was the idea of enrichment that led researchers to a vital discovery. Nutritionists knew that juvenile sea bass needed fats and proteins in their early diet. But if those fats and proteins were simply dumped into the water, juvenile sea bass would never be able to find them. It was the realization that a rotifer could be a perfect delivery system that proved critical. The very act of vibrating makes rotifers suitable "prey" for sea bass, compelling the sea bass to "hunt" and therefore acquire the fats and proteins that the rotifers contain. The final positive trait of rotifers is a rather strange quirk that makes them particularly useful to juvenile fish: rotifers possess an enzyme that causes them to literally digest themselves after death. This means that young sea bass, which early in life lack a full profile of digestive enzymes, can immediately get access to the nutrition contained within their miniature prey.

But rotifers were only the first phase of the solution. Once juvenile sea bass were over a few millimeters, they were still not quite ready for industrial feed pellets but were too large for rotifers to sustain them. A second transitional feed had to be used. And the one the international fish-farming community eventually settled on was one that, curiously, was more known to devotees of comic books than to readers of scientific journals.

Throughout the world, in the otherwise barren salt-lake ecosystems that occur in low-lying inland zones, a genus of tiny shrimp called artemia thrives. Because of the supersaline conditions of salt lakes, artemia produce hard, nearly impervious cysts that are fantastically tolerant to outside conditions. Artemia cysts over a million years old have been found and successfully hatched. The largest source of artemia in the world is the Great Salt Lake in Utah, and before marine aquaculture took off, there was an idiosyncratic mail-order marketer named Harold von Braunhut who popularized them.

Von Braunhut was a man of many weird talents. He raced motorcycles under the name "The Green Hornet" and managed a showman whose act consisted of diving forty feet into a children's wading pool filled with a foot of water. Originally of Jewish origin, he became a neo-Nazi and also invented X-Ray Specs—those red-threaded glasses that gave the wearer the impression of seeing through people's skin into their internal organs. But von Braunhut's most successful discovery was artemia. Since artemia eggs are so resistant to exterior conditions, von Braunhut reasoned they could easily be put in mail-order envelopes and sent around the world. All that was needed was a marketing name to make them attractive to consumers. He marketed them first in 1960 as "Instant Life" but in 1964 settled upon the name "Sea-Monkeys." Von Braunhut invented a whole parallel universe for these creatures—swings, playpens, life histories—all advertised on the backs of comic books. The hardy cysts could be mailed to eager nine-year-olds and would arrive ready to hatch in almost any water condition.

But European researchers realized that the best trick artemia could do was feed sea bass. The fact that they can be stored for years and then hatched at a predictable time makes artemia the ideal transitional feed for marine perciforms. Today the demand for artemia cysts is so great that overharvesting of them is now a major threat to the Great Salt Lake. One fish farmer I spoke with angrily compared the few nations who control the world's artemia supply to OPEC. In the last ten years, the price of artemia cysts has risen exponentially.

Both rotifers and artemia, though, have one quality that was causing Thanasis Frentzos's fish to be deformed. When they are enriched, they are literally overflowing with nutritious oils—oils that seep out of the animals' membranes and float to the surface of an aquaculture tank. Eventually Thanasis and his colleagues realized that it was these oils that were somehow interfering with the development of that key perciform organ the swim bladder. When the young fish were anesthetized for analysis, it was found that a good number of them simply sank to the bottom. Something was happening that wasn't allowing the fish to correctly form their flotation devices.

"We realized by carefully watching their behavior," Frentzos told me, that "at thirteen or fourteen days old, the fish would swim to the surface. And it was here that they would sip a tiny bubble of air. When the fish are this young, the connection between their mouths and their swim bladders is still open, and so they can put air into their swim bladder—the air bubble is what forms the swim bladder in the first place. But what was happening was that in the feeding environment we had created, the fish couldn't get to the surface. The oil from the feed was floating on the surface, preventing them from taking that first sip of air. And so we found that if we skimmed the oil off the top of the water, we would clear the way for them. Then they could take in their air properly before it was too late and they physiologically closed the barrier to the swim bladder."

All these developments would lead to greater and greater survival in the rearing of sea bass. Whereas in nature the survival of young was about one one-hundredth of a percent, by the time Frentzos had perfected his tank innovations, survival had risen to about 20 percent—a ten-thousand-fold improvement. And it was with these improvements that he was then able to make use of a uniquely Greek feature—the natural sea bass habitats of the Ionian Sea.

While France does have a relatively long coastline, environmentalists' resistance to near-shore fish farming combined with real-estate speculation meant that very few coastal sites were available for farming. Greece, on the other hand, is a natural for marine aquaculture. While Greece is ninety-sixth in the world in terms of overall land area, its shores are so intensely crimped and undulating that it is tenth in the world in total length of coast. No point on the mainland is farther than a hundred kilometers from the water, which means that fish grown near the shore can be easily and quickly shipped to major population centers.

Lastly, and perhaps most critically, because of the high degree of crenellation in its coast, Greece is endowed with bays and harbors with extremely low "fetch." Fetch is a nautical term meaning the distance over which wind-driven waves travel without encountering obstructions. The longer the fetch, the more powerful the waves, and the more powerful the waves, the more likely they are to destroy net cages suspended in the sea. Greece's thousands of bays are fetchproof wonders—surrounded by abruptly high mountains. Because fetch is for most intents and purposes not an issue, the Greeks could construct fish cages out of flimsy, found material at low cost. When marine aquaculture began, there were no advanced engineering schemes for building in-water fish cages, and they came up with the simplest solution; as one Greek aquaculture scientist put it, "Four floating blue barrels nailed together with planks, with a net hanging below."

And so it was in Greece that all the different elements came together. The very considerable problems of breeding, juvenile feed, and habitat had been overcome. The stage was set for the sea bass to go global.

The sea around Cephalonia is still largely empty of fish, and the fishermen who ply the waters live more off a 150-euro-per-day subsidy than off any of the fish they catch. But thanks to

Frentzos, many of the bays and inlets have net cages hanging in them today, filled to the brim with sea bass. The first crops grew out successfully, so successfully that local restaurants during tourist season were pestering Frentzos constantly for fish. "It got to the point where I had to hide the fish from others. They all wanted my fish!"

Word of Frentzos's success spread. Some of his staff were poached by other companies. And soon there was a Greek sea bass gold rush going on. Large sea bass–farming companies emerged that became publicly traded behemoths. So new was it all that the Greek government itself turned to Frentzos, who, while only slightly ahead of his colleagues, was asked to help set regulations on farming sea bass in the Mediterranean. "Once one of the farmers nearby came to me," Frentzos recalled, "and he asked me, 'Thanasis, can you please tell these people that the walkway between my cages is wide enough? It's sixty-eight centimeters and they're telling me it has to be seventy-five centimeters. And when I asked why, they said it's because you, Thanasis, have a walkway of seventy-five centimeters!'"

The other problem that stemmed from the sea bass boom was disease and pollution. Unaware of how to control disease proliferation in fish, newcomers to fish farming were overwhelmed by a bacterial infection called vibriosis. Frentzos was one of the only farmers who had proper training in marine biology and was more or less spared. He sited his farms in places with strong moving currents and maintained a carefully sterile environment. One farmer, thinking that Thanasis had some hidden knowledge that he could access, asked him for help. "I told him everything he had to do, but he wouldn't listen," Frentzos recalls, laughing. "He didn't do any of it. Eventually he hired a witch to remove the evil eye from his farm. Believe me, I understand the evil eye. Sometimes you feel it, watching you like the cat watches the mouse. But in this case it wasn't the evil eye. And this man, he went out of business."

But, like the French and the Israelis before him, Thanasis watched the market slip away from himself, too. During the ensuing years of the sea bass boom, being a good fish farmer with sound practices wasn't always the thing that made you successful. Greece, as one of the poorer nations to achieve European Union membership, had access to huge amounts of "cohesion funds" meant to bring struggling economies into parity with stronger nations like Germany. In Greece a large portion of those cohesion funds went to the farming of European sea bass. Again, the goal was not to help Greece produce a sustainable product—rather, it was to help Greece make money and enter the European economic zone with as little pain as possible. Less sound farmers than Thanasis could mask their bad practices when they occurred by just throwing more EU money at any problem that came their way. The most opportunistic of these Greek companies used EU money to build sea bass empires that have spread outside their borders. From a place of extreme scarcity, where sea bass were a dwindling source of wild marine protein, they have grown markedly every year in their farmed form. Today Greece sends nearly a hundred million of those exactly plate-size fish to diners throughout Europe, the United States, and beyond every single year.

Sea bass booms and busts now wash over the shores of the Mediterranean, rising and falling as profit margins get thinner and operations move to areas with weaker regulation and cheaper labor costs. From Turkey to Tunisia to Egypt they spread, and each time, as with the salmon industry in Canada and Norway, the same bad practices cause pollution, disease, and death. On each such occasion, emergency regulations are put in place to save the industry and the coast. Once upon a time, when sea bass were still rare, there was a profit margin of ten dollars per pound. Now that global production of sea bass approaches 200 million fish a year, the profit margin is down to half a cent per kilo.

The economics will only get worse. Feed costs are surging, and prices for sea bass are collapsing. The reputation of the European sea bass itself is also suffering. "I think both chefs and diners alike view it as a utilitarian fish," Jay Rayner, the food critic at the UK newspaper the *Observer*, wrote to me. "It stands up well to almost any accompaniment without over-dominating the plate." The holiday fish that was once a centerpiece of ancient Romans has become a day laborer that works to compete with side dishes. And the wages are not good. This makes Thanasis pull at his long white hair, wondering whether it makes sense to continue this strange adventure that he began in 1982. The world needs food, he insists, and the people who are hiking up costs on him are not thinking wisely about the future.

"When these accountants from these companies call me and tell me they are raising their prices on me again, I look forward to the day when they will have nothing to eat. That will be some kind of day. And you know what I'll say at that point? I'll say, 'Eat your computer. Eat it. Eat it right now.'"

Socrates," I asked, "how do you tell a wild sea bass from a farmed sea bass?" I was at a restaurant northeast of Athens with representatives of Selonda, the second-largest sea bass farmer in Greece. A grilled, whole European sea bass lay on a plate before us. Socrates

Panopoulos, a hatchery manager of the Greek-owned company, let his junior scientists answer first.

"The liver," one biologist proposed. "If the liver is dark red, it means the fish is eating a low-fat diet and is probably wild."

"But, Kostas," Socrates replied, "they have taken out the guts. What are we to do?"

"Maybe the otolith?" another scientist suggested.

"Ah," said Socrates, "the otolith."

He worked a knife into the fish's head and removed the pearly ear bone. Then, using his glass as a primitive magnifier, he counted the otolith's layers, which accrue like rings in a tree. "I see four," Socrates said, "and they are uneven. This fish is four years old and wild."

I had purchased this wild sea bass, to test the Selonda scientists' know-how, at the Central Fish Market of Athens, where one can still find a separate display for "wild" sea bass and where such wild fish sell at three to five times the price of their farmed equivalents. Whereas there was no such thing as a farmed sea bass forty years ago, today there are nearly ten times as many sea bass grown as caught. Around 180 million pounds of farmed sea bass are put on ice yearly, compared to just under 10 million pounds of wild fish. But as Yonathan Zohar pointed out to me on my return from Greece, even the supposedly wild fish may be radically less wild than they were half a century ago.

"There is no doubt in my mind," Zohar wrote me from his lab in Baltimore, "that the so-called wild sea bass are escapees from the cages. We can confirm this via DNA tests." Whereas three genetically distinct populations of sea bass once inhabited the Mediterranean—an Eastern, Western, and an Atlantic stock—now the Western stock, developed primarily by the French, probably predominates, even outside fish farms. "Fish of Greece, Fish of the Sun" may be the motto under which Greek sea bass are sold to the world, but these fish are, genetically anyway, fish of France.

The taming of the sea bass has brought both good and bad things to the world. Indeed it needs to be held up for analysis as we chart the way forward. By the standard metrics of domestication, the sea bass was not the best choice to be the first ocean perciform in our mangers. It is difficult to breed, it is hard to nurture past its larval stages, and it requires more wild fish as feed than it ultimately yields at harvest. If we were truly desperate to come up with a better source of food to "feed the world," we would have chosen something else.

But look at the environment in which the project was conceived. The reasons the Israelis, the French, the Italians, and the Greeks chose the sea bass were as much based on profit as they were on finding something to alleviate hunger. When they set out to tame the sea bass, their goal was not to turn a holiday fish into a weekday fish; rather, in the fashion of the Green Revolution fantasies of the day, they imagined the fish as part of a year-round holiday. When the idea of farming sea bass was first hatched, no one could have foreseen that they would be so successful as to cause a massive collapse in price and return on investment.

We must also keep in mind that the European sea bass was the Rosetta stone of fish—the animal that unlocked the secrets of development for every major commercial species of ocean fish in the world. Its cuneiform is the shape of hormonal chemicals, the dietary needs of juveniles, the light intensity of the winter sun—all the factors that were once the exclusive secrets of the primeval Mediterranean salt marsh during the halcyon days. Secrets scores of millions of years old. And these secrets turn out to be strangely common to the vast order Perciformes. With some modifications here and there, the sea bass template of breeding and rearing seems to work for most of the fish we eat.

And now remains the question of what to do with those secrets. Do we use them to decode more and more species and create a whole parallel domesticated world of fish, a man-made sea within a wild one, exclusively for human use, regardless of what effect it might have on a wild population? Or do we develop species and farming methodologies that take into consideration both their impact on the natural world and their benefit to humanity? The domestication of any animal is always difficult and brings with it a host of complications and potential diseases. It should never be taken lightly.

If there is one person who has taken in all the lessons of the Israelis, the French, the Italians, and the Greeks in their quest to tame the European sea bass over the course of the last quarter century, it is Josh Goldman, an earnest, limpid-eyed fish farmer in his mid-forties who installed himself near his old college town in the bucolic Pioneer Valley of western Massachusetts. Not a fish person by habit, Goldman came to fish farming through his training as an ecologist. He saw in fish farming a great ecological promise. Ecological theorists posited in the 1970s that aquaculture, done right, had the potential to achieve the most elusive goal of animal husbandry—to produce one pound of flesh for less than a pound of feed. Because they don't swim against gravity or raise their body temperatures, fish require substantially less energy than do land animals. If done right, farmed fish, food scientists believed, could solve the world's protein problems with a snap of the fingers.

But Goldman watched in dismay throughout the 1980s and '90s as, one by one, first with salmon and then with the European sea bass, aquaculture sullied its reputation and became cast in the public mind as a dirty industry with a net protein loss for the world. European sea bass, like salmon, require as much as three pounds of feed for every pound of flesh they produce.

It was in trying to right this equation that Goldman decided to invert the processes of artificial selection. Instead of looking for a known and popular fish, a holiday fish, and trying to tame it and make money off its reputation, Goldman decided to find a fish that was a natural partner for humans. "I remember I was in a used-book store back in the early 1990s, and I came across a book by the legendary British fisheries genetic scientist Colin Purdom," Goldman told me on a crisp fall day, the leaves on the trees almost palpably turning to crimson through the window behind him. "Purdom made this argument. He said before you start a domestication program, you need to understand what nature has been up to, starting with as large a selection from the wild as possible in order to understand and hopefully benefit from the enormous diversity that's out there. I realized we were spending all this money trying to solve some really difficult reproductive and breeding problems. But in our focus on these issues we never asked fundamental questions about the species itself before choosing it. What does it eat? What are its fundamental behaviors? How fast does it grow?"

These questions launched Goldman on an epic global quest. He abandoned an early project with the striped bass that he had been raising because he found that they failed two of Galton's criteria.

1. They did not breed freely—they spawned only once a year around the April full moon and are very difficult to breed in captivity.

2. They hated being handled by humans, thrashing in

their nets, shredding their scales, and often fatally wounding themselves.

Goldman tested over fifty different species, looking further and further afield, each time finding some fatal flaw in the fish he investigated. But finally, at the dawn of the new millennium, he met an energetic Australian entrepreneur named Stewart Graham, who introduced him to a Southeast Asian fish that met all of Goldman's criteria. Colonial-era Britons had named the fish the Asian sea bass, but it was even more distant from the European sea bass genetically than any of the other American basses.

The fish was known locally in Australia by its Aboriginal name, barramundi. And its natural environment was almost a replication of what humans typically create when they make fish farms. Whereas sea bass and striped bass spawn in fresh water and live their adult lives in the open saltwater sea, barramundi are catadromous and do the exact opposite. Living this way means they migrate into fresh water in summer and often get stuck in billabongs, areas of rivers that get cut off and become stagnant during dry seasons. In an odd way what billabongs resemble most are big, natural aquaculture tanks. Whereas the striped bass thrashes like a banshee when confined, the barramundi is docile and compliant and takes well to handling. Barramundi are, under the right conditions, wildly fertile, spawning throughout the year. They have adapted huge gills for anoxic environments, which makes them highly disease-resistant. Finally, and most important, they can live mostly on vegetarian feed, which means they are less reliant on fish meal and oil than are other carnivorous species. They are also therefore less prone to contamination from tainted feed—the less fish feed in an animal's diet, the lower the chances that the harvested fish will have unacceptable PCB contamination. At the same time, barramundi can do something almost no other fish can do—they can make the omega-3 fatty acids from vegetarian oils.

"Aquaculture's promise and its central challenge," Goldman told me, "is to dramatically increase the world's fish supply while using an essentially fixed amount of wild fish in feed. Given the projections that fish farms will need to double their output to keep pace with demand over the next ten years, there is a pressing need to make far better use of these limited natural ingredients so that we don't continue to drain the world's oceans in the process of trying to keep pace with demand."

Again, it is the barramundi's natural history and life cycle that make the fish a potential net marine protein gain for the world. Because barramundi live in fresh water and spawn in salt water, they must quickly synthesize enough omega-3s to pass on to their eggs when they migrate to the sea and spawn. In aquaculture only a small amount of fish oil and meal needs to be given to barramundis as part of a "finishing diet," as it is called in the industry.

But Goldman would never have been able to bring the barramundi into a tamed environment without the groundwork laid by the European sea bass pioneers. Without the live-feed techniques of the French and the Dutch, they would have starved as juveniles. Without the hormone implants developed by Zohar and the Israelis, the barramundi's breeding would never have been completely regularized to allow a steady, consistent crop. Without the scaling-up of the Greeks, Goldman would never have been able to imagine the advantages of large-scale production.

And it *is* large-scale production that he's going for. In Turners

Falls he and his partners have created the biggest "recirculating aquaculture system" in the world. These Asian barramundi, a most exotic species, never have any contact with the living ocean. Dozens of house-size tanks in four different airplane-hangar-size rooms (each named after a different Australian city—Adelaide, Brisbane, Canberra, and Darwin) draw water from wells fed by the Connecticut River, constantly recirculating and cleaning it, making a completely sterile environment where the fish grow fast and seldom suffer disease.

So at Turners Falls, Massachusetts, the site of one of the more tragic fish extirpations in contemporary history, the place where Connecticut River salmon were wiped off the face of the earth with a single dam, the possible reinvention of fish as food as we know it could be happening. An animal has been chosen specifically for its small ecological footprint and its natural tendency to adapt to human culture. A fish that could fulfill all the Green Revolution promises of the early days of ocean aquaculture. One that would truly generate more fish for the world than it would consume. One that could actually take pressure off wild stocks of similar species and cause humans to lessen their impact on the ocean overall.

There's only one problem. Unfortunately, no one in the West seems to know what a barramundi is. Goldman has recently taken his eco-friendly recirculating fish farming technology to central Vietnam to produce frozen barramundi for the U.S. and European markets. In Asia, production of the fish is increasing. Around 90 million pounds of it are grown annually, not just by Goldman but by farmers in China, Australia, Indonesia, and elsewhere. In Europe and the United States, though, most consumers remain reliant on aquaculture species that eat too much and carry too large an environmental burden.

Perhaps Goldman should take one more page from the history of the Mediterranean and look for a name that people will recognize. Knowing what I know now about fish—how they brand themselves, how humans come to accept them—I might call the barramundi something else. I might call it by the name that colonial Britons imposed. A name that for whatever reason has made its way into our consciousness as indicative of a "good" fish. Why not call it "Asian sea bass"?

Cod

The Return of the Commoner

The transformation of salmon and sea bass from kingly and holiday wild fish into everyday farmed variants is a trend that continues with different animals around the globe. Using the technologies developed for the domestication of the European sea bass, many other high-value species that typically fetch in excess of fifteen dollars a pound—like sturgeon, grouper, and even bluefin tuna, as we shall see later—are all at various stages of being tamed. Ultimately, though, these are niche fish for niche markets, developed, at least initially, to compensate for local population declines or extirpations that occurred during the first big local fish crashes of the 1970s and '80s. But what do you do when you start to lose not the holiday fish but the workaday fish, the fish upon which average people rely for their daily meal, the one that should sell at around the same price as chicken? The fish whose very abundance is its most notable characteristic?

In the spring of the year 2000, a book that spoke to these questions started working its way around my family's small circle, passing first from my brother, who had brought it over from England, on to me and then to my aunt and uncle. The book was by a onetime commercial fisherman–turned–journalist named Mark Kurlansky and was called simply *Cod*. It is considered to be the first of what the publishing industry would come to call "the microhistory," in which human social evolution is traced through a single commodity. In Kurlansky's case that thing was *Gadus morhua*, or the Atlantic cod, a species whose flaky white flesh nourished humanity from medieval times through the discovery of the Americas and on into the industrial era. If the European sea bass is the epitome of the specialty fish, then cod, as Kurlansky made clear, represented the opposite: sheer abundance and commonness. Enough abundance to help grow the human population of the Western world twentyfold.

Different members of my family appreciated different elements of the book *Cod*. My aunt liked how Kurlansky had delved into medieval cookbooks and reproduced weird recipes for cod livers and swim bladders, or "sounds." My brother (an early Dungeons & Dragons adopter and Yale medieval-studies major) enjoyed the bit about how Basque fishermen may have discovered America long before Columbus, but kept it a secret because they didn't want to share their good cod-fishing spot with other nations.

But for me, I fixated on the story of the loss of abundance that lay at the center of the book. The loss of abundance and the greedy privatization, monopolization, and industrialization of fishing that caused it. How, after a series of technological advancements, huge corporations overwhelmed smaller artisanal fishing fleets with giant factory ships and obliterated one population of cod after another. Just as the Italians had turned to dynamite when sea bass stocks became thin in the Mediterranean, industrial fishing fleets—first in the North Sea, then off Iceland and over to the shoals off Nova Scotia and on down south into New England—got progressively more destructive.

Large bottom-dragging nets, or trawls, grew larger and larger and became equipped with "rock hopper" attachments that allowed factory ships to penetrate into the last craggy redoubts of codfish banks—offshore upwellings of fertility where codfish congregated to do their annual mating and spawning. And all along, fishing efforts grew with the support of scientists who claimed that codfish populations could not be overfished—that overfishing wasn't really a provable scientific concept. The heavy fishing was ramped up even more when the U.S. government dumped $800 million in subsidies to build up the American fishing fleet in the late 1970s and early '80s.

Cod concludes with the federal government's finally closing the greatest New England fishing grounds, Georges Bank, to commercial cod fishing in 1994, the very grounds that gave Cape Cod its name. "Is it really all over?" Kurlansky lamented of the dry-docked Massachusetts cod fishermen at the conclusion of his moving, epic book. "Are these last gatherers of food from the wild to be phased out? Is this the last of wild food? Is our last physical tie to untamed nature to become an obscure delicacy like the occasional pheasant?"

These words stayed with me over the years to come. But histories of environmental wrongdoing have a strange way of putting traumatic events in the past, sealing off bad human behavior of former times from the unwritten pages of the present and the future. When such books become popular, there's the impression that the various transgressions they expose have reached the public and that things must inevitably have changed for the better. Just as Rachel

Carson's *Silent Spring* had compelled the U.S. government to ban the pesticide DDT and help eagles, falcons, and hawks back to viability, I hoped that *Cod,* an international bestseller on an order of magnitude that no other fish book had achieved since *Jaws,* had brought the question of overfishing into public consciousness. Kurlansky's book was published in 1998. By the winter of 2008, I wanted to believe that things had changed, that all the abuses that had been heaped on cod had been studied carefully by the scientific community and cleaned up by way of effective policy.

Moreover, there were signs in other fisheries that the hopeless spiral of decline and extinction could be reversed. The American striped bass, which also had suffered a tremendous collapse in the late 1970s, had been protected by a three-year fishing moratorium in 1982 followed by an extreme reduction in commercial-fishing pressure thereafter. Today there are more striped bass around than at any time in the last hundred years. Surely, I reasoned, after fourteen years of commercial-fishing closure, conservation activism, and scientific research, it must be possible to go fishing for cod again on Georges Bank.

Logistically, anyway, it was. Portions of Georges Bank have been opened periodically to fishing over the last two decades, and the party fishing boat *Helen H* out of Hyannis, Massachusetts, it turned out, was making daily trips to the bit of fishing ground around the banks that was open to recreational fishing. The December after I returned from my bass research in Greece, I received special dispensation from my family for a full twenty-four hours' sea leave to give it a shot. Hitting the road from Manhattan late on a Friday evening, I drove through the night up I-95, the highway that had been my thoroughfare to many of the spots I'd fished along Long Island Sound in my youth.

After New Haven the long urban reach of New York City started to fade and the highway stayed mostly dark, with the exception of a burst of light as I passed by my old college town of Providence, Rhode Island. Then on I-195, a little while after crossing into Massachusetts, the beautifully ominous Bourne Bridge loomed. The Bourne Bridge reunited Cape Cod to the mainland in 1933 after the Army Corps of Engineers had cut the Cape off with a canal in 1916. It was quieter than quiet now, the cold seeping into my car through the iced-over windows. On the other side of the bridge, with a light snow starting to fall, I drove a few more miles to the old whaling town of Hyannis and at 2:00 A.M. pulled into the *Helen H*'s parking lot. There, much to my dismay, was a huge crowd of unruly New Yorkers who had just made the exact same journey.

They were bunched up in groups, according to nationality. Over to my left, with ice chests the size of coffins piled up on a kind of latter-day oxcart, were fifteen or so Koreans. To my right, a dozen Dominicans stamped their feet against the cold and rubbed their hands together in anticipation. Greeks, Croats, and other assorted seafaring ethnicities were also gathered up in smaller numbers.

Still, even with the crowd and the long journey, I was glad to be in Hyannis that cold morning and glad to see a party boat with a full load in pursuit of cod. The American party fishing boat represents to me something special in the world—it is an outright acknowledgment of abundance. Unlike private charter boats, which can charge well over a thousand dollars for a day's fishing, party boats are big two-hundred-foot bruisers that can hold sixty, even a hundred guys (and yes, they're usually guys) and often make the trip for fifty bucks a man. The existence of a fleet of these kinds of ships that charge a reasonable sum to blue-collar folks implies that wild fisheries are still a common, reasonably plentiful, and useful resource.

We all queued up against the shed to pay our fare. I recognized some of the people in line from the party boats I fish in summertime out of Sheepshead Bay, Brooklyn. Not wealthy people by any means: plumbers, carpenters, import-export stevedores. But then again, who was I to speak of wealth? Every single magazine and newspaper I wrote for teetered on the edge of extinction. The idea of deficits and cutbacks was on everybody's mind. And the huge crowd at the fishing boat was not your average bunch of "recreational" fishermen. They were fishing for meat. Indeed, my own rationalization for making this trip had been in part economic. Since this was an offshore run, nearly three times as long as a Brooklyn day-boat trip, the fare was higher than usual—$170. The associated expenses of gas, tolls, and a gratuity for the mates would bring the total cost to something like $250. But the reports coming out of Hyannis had been good. Since cod populations had collapsed, the price of cod had risen quickly—the fish now fetched $13 per pound. I would need to bring back about twenty pounds of fillets—ten good-size fish—in order to break even. Any more and I would be beating the market.

Fares were paid, and then the captain appeared and did the roll call, boarding passengers in order of reservation. The boat was sold out, and I had reserved the last available place over the telephone. I feared I would get a horrible spot near the bowsprit. But when I was called at last, I pushed my way through the crowd and spotted, miraculously, a widowed stern slot just off the corner, the place where, in my experience anyway, the greatest amount of fish are caught. My fishing place claimed, I found the last free bunk, and the *Helen H* began a slow grind seventy miles east toward the Nantucket Shoals, the slope leading to the underwater upwelling of Georges Bank.

Just before dawn the engines slowed and the swish of rubber coveralls being slid over tired bodies woke me. I rushed out of my bunk to do the same. The *Helen H* circled, its sonar running, trying to locate a concentration of fish big enough to satisfy a full boat. When the engine finally cut out, the captain called us to the rails and the fishing began. Only a slow trickle of fish came aboard at first, and I remembered the voice of the *Helen H* secretary when I'd called to ask how the fishing was. "Excellent," she'd said, almost too definitively. Was that the telltale hint of professional exaggeration in her voice?

Fishing next to me, the Koreans, always good fishermen, had the first catches of the day, but they were unimpressive. One man on my left reeled in. "Small size," he complained, holding up a codfish short of the twenty-two-inch federal size limit, then throwing it into his cooler anyway, despite the law. "Need a keeper size!" he said, as if to somehow excuse the fact that he had just killed an illegal fish by expressing that he *wanted* a keeper-size fish. To my right, another man fished with three hooks on his line (federal law allows only two), and one person over was another guy pushing the limits of legality by using an electric reel. Every time his lure hit bottom, he would immediately have a fish on. He'd hit the "retrieve" button on his reel, and with a loud zzz-zzz-zzz the reel would wind effortlessly up and he would throw an illegally short cod into his bucket.

I, too, was catching small fish, but I treated them with excessive kindness when I brought them on board, keeping my fingers out of their gills, touching them as little as possible in order not to disturb the slimy mucus coating that envelops most fish and is their primary defense against topical infection. One fish I caught got snagged through its gut, and even though it would surely die, I returned it just the same, obediently following the rules. I caught a single keeper-size fish during the first hour and moodily stared at it in my cooler. "The two-hundred-and-fifty-dollar fish," my father would have said. What the hell had I made this trip for?

At a certain point, though, the boat seemed to settle into its drifts and we somehow became more and more attuned to the size and movement of the body of fish below. The cod on our lines inched up in size, some of them well past the legal limit—a limit that had been imposed to ensure that as many codfish as possible got at least one successful spawning before being caught and killed. And soon the fishing became downright ridiculous. A dozen cod were now in my neighbor's cooler and about half that many in mine.

Now when I felt the tug of a small fish on my line, I left the jig down on the bottom to ensure that a second, larger codfish would hit the second, "teaser" hook. As the fishing got better and better, I became increasingly disrespectful to the short fish I had so pampered at the beginning of the day. Two small fish on my line meant the double task of removing hooks on unprofitable fish so that I could get back down to the bottom more quickly and catch more keepers. Instead of gently sliding hooks out of jaws, I ripped and pulled and put my fingers in gills to gain a better purchase.

In the back of my mind, I remembered something Mark Kurlansky had mentioned in *Cod*. Something about how abundance with fish breeds, if not contempt, certainly a diminishing of respect.

But that was just background noise. The abundance appeared all around the boat, in every incarnation. Black pilot whales crested in the smooth sea. Now and again, hundred-pound bluefin tuna, which I had been led to believe were nearing endangered status, cut the water with their crescent dorsal fins and exploded into the air in pursuit of prey. And the prey! Sand eels and herring by the millions dimpled the water and found themselves impaled on our jigs when a codfish did not occupy them. When one particularly large codfish came to the surface, I had to hurry to bring it in as an eight-foot porbeagle shark emerged and threatened to bite my catch in half.

Soon the codfish became too numerous for my cooler, and I ordered up a second tote. When the captain blew the horn three times, signaling the end of fishing, I had nearly two dozen cod in all, my best day of cod fishing in many, many years. I had the mate fillet up all except two, which I left intact to show my two-year-old son, in the hopes that it would inspire him to become my fishing buddy in my later years. Lifting the cooler up and placing it into storage for the long ride back to port, I judged its weight in proportion to how my son feels in my arms. When he was born in 2006, he was twenty-two inches and six pounds, the exact size of a keeper codfish. At the time of my fishing trip he weighed thirty pounds, and my cooler felt twice that weight. I had surpassed my economic goal. I had spent a total of $246 but had managed to procure sixty pounds of codfish fillets, giving me a per-pound cost of just $4.10.

A wild product more natural than organic and as cheap as chicken breasts. I climbed up into my sleeping bunk and tried to get some rest. And though the mutterings of the crew members could be heard below about how it had been a "so-so day" and "nothing like before," the ache in my arms and the thought of my full cooler indicated a blissful, almost dizzying abundance.

"Take that, Mark Kurlansky!" I said to myself as I drifted off to sleep.

How many fish do we need for our food, and how many are there in the sea? Ultimately, all other questions about the future of the ocean fade in importance compared to these two essential inquiries. If we cannot figure out this basic equation, the continuation of marine life as we know it, or at least marine life as we want it to be, will be impossible.

With respect to question number one—how many fish do we need?—it is possible to come up with a very rough estimate. Currently the world's wild catch measures 170 billion pounds—the equivalent in weight to the entire human population of China, scooped up and sliced, sautéed, poached, baked, and deep-fried, year in and year out, every single year. This is a lot of fish—six times greater than the amount of fish we took from the ocean half a century ago. But if we were to follow the advice of nutritionists, the amount would be far greater. The British Department of Health, for example, suggests that a person should eat a minimum of two servings of fish per week— one serving of oily fish, like salmon, and one serving of whitefish, like cod. So if every single human being were to do what the British government says we should do, we would require 230 billion pounds a year—60 billion pounds of fish more than we harvest at the most exploitative period in history.

This, then, puts a lot of pressure on the second question—how many fish are there? Finding the answer is much trickier. No one really quite knows how many fish there are and, more important for the future, how many fish there could be. The United Nations' Food and Agriculture Organization (FAO), a kind of database of databases that compiles all the fisheries information from all the governments in the world, wrote in its most recent assessment of global fish stocks that "the overall state of exploitation of the world's marine fishery resources has tended to remain relatively stable. . . . Over the past 10–15 years, the proportion of overexploited and depleted stocks has remained unchanged."

And yet this assessment of stability is up for question. The FAO fact-checks and audits its findings but admits that despite its efforts, its "fishery data are not fully reliable." The FAO also notes a "spreading of over-intensive fishing from the northern to the southern hemisphere" and has provided "a consistent warning . . . about the consequences [of overfishing] for the overall sustainability of the global fishery system."

Nowhere is the stress on abundance more acute than with cod and the other whitefish that are used as "industrial fish"—the raw material for fast food and frozen supermarket meals. Whitefish today represent about a fifth of the world catch, or the whole human weight of the United States. They are principally animals of the order Gadiformes, cod being the most famous example of that order. But haddock, hake, pollock, and other gadiforms are increasingly folded into the mix. In Northern European countries, particularly the United Kingdom, cod and other whitefish represent the most-consumed fish in the national diet, making up nearly a third of all the seafood Brits eat, both as cheap fast food and as more expensive fresh fillets.

The popularity of gadiforms is greatly aided by their morphology generally. Gadiforms live a lazy life, preferring to move slowly in chilly water. Their flesh therefore often contains a minimum of high-speed muscle tissue—tissue that is usually contained within the blood line that runs down the length of the fish's fillet. Because the size of a fish's bloodline contributes to its "fishy" flavor, cod do not taste very fishy. Cod also have a tendency to store oil in their liver rather than in their flesh. Since oil in the flesh determines the speed at which flesh putrifies when frozen or dried, cod and other gadiforms can be stored for great lengths of time. Gadiforms are therefore the perfect industrial fish: they are common, mild, and easily recast as different kinds of food products. Whether used as dried bacalao to feed slaves on southern plantations in the nineteenth century or as fish sticks to feed working families in the modern era, gadiforms should be—in a healthy marine ecosystem anyway—so plentiful that they shouldn't seem special in the least. They are the true everyman's fish.

Gadiforms' wide use as industrial fish also stems from the fact that they are available in waters of *both* hemispheres. Generally, animals that evolve in a cold-water ecosystem end up confined to a single half of the globe, because once they have adapted to cold conditions, warm equatorial climates are deadly and act as a kind of prison wall, effectively sealing off passage from one pole to another. It is for this reason that penguins are found only in the Southern Hemisphere and puffins only in the North.

Gadiforms, though, are older than the hemispheres as we know them. The ur-gadiform is thought to be something found in fossil form in the North Sea, called *Sphenocephalus*. Modern gadiforms evolved from the extinct genus *Sphenocephalus* and radiated out to the Southern Hemisphere at a time when the continents were more unified and a bridge of frigid currents allowed cold-water fish to cross the equator. But approximately 45 million years ago, Australia broke away from Antarctica during the Eocene epoch, and a powerful whirl of water took shape that whipped around newly isolated Antarctica. This shunt, called the Circum-Antarctic Current, continues to spin at the bottom of the world today; it effectively walled off large populations of Southern gadiforms from their Northern cousins. These Southern gadiforms are only now coming under intense exploitation, but more about that later.

Over the eons certain gadiforms, cod being the best example, developed an approach to life that grabbed every opportunity to become more numerous and then in turn used that abundance to perpetuate abundance over time. More sedentary fish that do not migrate broadly must content themselves with the biological energy available in their immediate environment. The sun can transmit only so much energy into a given square mile of sea, and that energy in turn can result in only a certain amount of plankton calories that are in turn passed up the food chain. But codfish travel far and wide and are highly omnivorous. Before human intervention the many eco-systems they traversed allowed them to build up extremely large populations.

Whenever there is food energy available in a system, there is competition from multiple species for that energy. Cod bucked this competition by outright numerical dominance, managing over time to monopolize much of the energy pathways of the North Atlantic. The most analogous example to cod would be the Douglas fir, cedar, and redwood forests that came to dominate over other plants in the Pacific Northwest. Just as tens of millions of huge redwoods and Douglas fir trees spanned western North America from San Francisco to British Columbia and formed dense ceilings that blocked out all light and suppressed other tree species, so, too, did the vast schools of cod form a kind of predatory canopy over the continental shelves around North America and Europe. On Canada's Grand Banks, codfish regularly reached five feet in length and weighed upward of a hundred pounds. Crabs, lobsters, mackerel, and other creatures that might have preyed on smaller, more vulnerable cod when they were first hatched were kept at low levels by the gape-mouthed, marauding hordes of big bad cod that monopolized the most productive swaths of current.

Of course, with the advent of industrial factory fishing, a different kind of ecology was established, one where the climax community that came to dominate the continental shelf was not a fish of the order Gadiformes but rather a hominid of the order Primates. And though there is no 100 percent provable scientific correlation between the rise of man and the fall of fish, it is instructive to muse on the inversion. According to Jeff Hutchings, who studies historical cod abundance at Dalhousie University in Nova Scotia, the total cod population of Canada's most famous grounds was even greater than that of the American buffalo, and the decline of cod represents the greatest loss of a vertebrate in Canadian history—somewhere on the order of 2 billion individuals. "What's the equivalent loss in human weight of two billion cod?" Hutchings wondered aloud. "That's twenty-seven million people." After speaking with Hutchings, I cast around for a nation that had a population of around 27 million people—fishing for a good metaphor, so to speak. Ironically, a country that has a pretty close equivalent in human biomass to the lost Canadian cod biomass is Canada.

What happens to a climax marine ecosystem, then, when the climax animal is taken from the water and turned into seafood? With fisheries the result is dependent on the extent of the removal. Hutchings's research has found that a removal of 70 percent to 80 percent of a fish population has a certain degree of reversibility. In a case where 20 percent or 30 percent of fish are still in the water, the population may be unstable and vulnerable but still has reasonable potential for recovery. Also, the genome of the stock—that is, the sum of the genetic diversity of the population as a whole—is not likely to have been heavily depleted; there are still nearly as many genes that account for a healthy amount of variability that will enable population recovery over time.

But when removals slip upward to 90 percent or more, the chances of recovery diminish and it is possible that the genome itself may be affected. In the fifteen years since the Grand Banks were closed, where total removals exceeded 95 percent of the historical cod population, a noted decrease in the size of the average codfish has been observed. Instead of averaging twenty-odd pounds as they did when the climax community was still in place in the 1960s, the average cod is now about three pounds—"scrod" size, as fish marketers like to call cod that produce a pan-size fillet. Furthermore, even after the Canadian government instituted a total closure of the Grand Banks in 1992 followed by severely restricted fishing later in the decade, codfish abundance has not shown signs of significant increase.

What this suggests is that by catching all the big cod, fishermen have in a sense selected for small cod. The genes for small cod may now be more frequent in the overall cod genome than they were before fishing pressure was applied. Even if cod on the Grand Banks were left alone, it might take many decades for the population to recover its previous genetic potential and reclaim the average size required for dominance.

The collapse of the Grand Banks cod was, for the years preceding Mark Kurlansky's *Cod,* a local problem of the Canadian Maritime Provinces. It prompted hand-wringing on the part of Canadian officials and would get coverage on occasion from the national Canadian press and from newspapers in the cod-eating countries of Northern Europe. But when the cod crisis spread to the United States, the issue became a global metaphor. The loss of abundance, when it happens in front of your face, is a shocking thing, especially when that thing is such an important food source to so many working-class people.

But cod present an interesting twist on the old model of threatened-species recovery. Both humans and codfish require *extreme abundance* of codfish for their continued prosperity. At the world's present rate of consumption, humanity needs about 40 billion pounds of codlike fish annually, more or less the size of the entire Grand Banks codfish population at its highest recorded level ever, *every single year.* And by the late 1990s, it had already become apparent that some drastic measures had to be taken to try to restore the abundance of cod. In the period between 1962, when codfish stocks were still healthy and menu items like McDonald's Filet-O-Fish sandwich were invented and produced at a cost of less than twenty-five cents a sandwich, and 1994, when most cod stocks were considered "collapsed" (collapsed generally being defined as a state where 90 percent or more of the historical population is gone), the United States, Canada, and the United Kingdom all went from being net exporters to net importers of not just cod but all fish.

The problem then becomes much more complex than dealing with an endangered species like, say, wolves, where there is no longer any harvest pressure. There may be arguments about how many wolves we want, but there is no "wolf industry," waiting, guns in hand, to pick them off if the population reaches harvestable size. How do you restore an ecology of abundance when even the diminished system is still being plundered by humans? And how do you do this when you must reestablish the dominance of a fish that has been not just turned into a second-stringer but genetically demoted into a fragment of its former self? A tough job indeed.

But not impossible.

To begin with, it is important to note that even though all cod populations in the world are technically a single species that can interbreed, different populations of cod have different prospects for the future, depending on how much they have been exploited and how well they are reproducing. You cannot say, "Cod are endangered because the Grand Banks cod have collapsed," any more than you can say, "Humanity is starving because the Sudanese don't have enough to eat." Correspondingly, there is no "global cod solution" that will save every cod stock in the world.

There are, however, basic human mistakes, fundamental misapprehensions of the potential of natural systems that point out just how primitive and ignorant we continue to be in our relationship with wild fish. And as it happens, one of the most important places where this ignorance was brought under the bright light of rational science was Georges Bank, the place from where I had extracted two dozen codfish in 2008 and where Mark Kurlansky concluded his book about cod in 1998.

Overfishing" is a term that is used widely in the press today. Those who follow fish have the impression that it is a stable and accepted concept within the scientific community. But even though the ocean clearly has some kind of limit to the total amount of fish that can be removed, until relatively recently no government agency was willing to go on record as defining what overfishing meant in general and which fish specifically were overfished. It was only in the 1990s, while scientists and policy makers were hashing over the question of Georges Bank codfish, that the concept of overfishing finally stepped into the ring to fight against the forces of ignorance.

Professor Andy Rosenberg, an ecologist at the University of New Hampshire, had a ringside seat to the fight. Though he started out as someone destined for a place up in the bleachers of academia, he got his Ph.D. in Nova Scotia just as the Grand Banks crisis was unfolding. The lessons he learned there prompted him to apply to be part of the American fisheries management system just as the American cod crisis was coming to a head. Fisheries in the United States are regulated by a series of regional management councils (FMCs, in fisheries parlance) that are by law a mix of representatives of the fishing industry and qualified scientists. Historically, it is the fishermen who have decided what is and isn't "sustainable," and science generally served as a means of supporting fishermen's claims. But in 1994, when Rosenberg applied for (and to his surprise was given) the directorship of the FMC that oversaw New England cod just as New England cod were about to plummet into extirpation, the rules of the game had begun to change.

"By the beginning of the 1990s, it was pretty clear there had been some good codfish reproduction in 1989. And to a lot of people it looked like the Georges Bank codfish's last gasp," Rosenberg told me. "The advice from the scientific community was that you really had an opportunity with this last good burst of fish to take some strong management action and put the fishery on stable ground." Action, however, wasn't taken. Not until a nonprofit organization called the Conservation Wildlife Fund sued the federal government for not fulfilling its duty to protect fish stocks. "And at this point the government did something extraordinary," Rosenberg recalls. "They said, 'You're right, we haven't met our responsibility. Our own science agrees.'"

Leading this leap into rationality was a surprisingly progressive director of the National Marine Fisheries Service, named Bill Fox. He would later be vilified by the fishing community for taking the side of science rather than the side of fishermen. But his act would eventually improve things for fish and fishermen alike. In 1993, for the first time in perhaps the entire history of the world, Fox required fisheries managers to define overfishing and to stick to that definition in planning fishing for the future. By the time Andy Rosenberg took up his position in 1994, it had been concluded by the newly science-driven Northeastern FMC that the only way to stop over-fishing on Georges Bank was to stop fishing altogether. And in 1994 this is exactly what happened. Two huge swaths of the banks, areas that were considered the best places to fish in the whole Northeast, were closed. The measure was considered temporary at the time, and many are still waiting for the closures to end.

But think about it. What happened in 1994 in Georges Bank was something completely new in the history of man and fish: the United States had created a de facto marine reserve in the middle of one of the most exploited fishing grounds in the world. This would eventually have much broader implications. The New England cod crisis would prove to be just the inciting incident of a larger move-ment, a movement that led to a landmark piece of fisheries legisla-tion known as the Sustainable Fisheries Act.

Before the Sustainable Fisheries Act, the default assumption about the ocean had been that it was inherently abundant. While today the total harvestable catch of the oceans is put at about 90 million tons, as recently as the 1970s some in the scientific com-munity suggested that humans could potentially harvest 450 million tons of seafood every year—or about the entire weight of the current human population of the world. This was reflected in fisheries leg-islation. As Michael Weber reported in his excellent book *From Abundance to Scarcity,* prior to the Sustainable Fisheries Act the bur-den of proof was put on conservationists to prove that a given stock of fish *wasn't* abundant enough to support a commercial fishery. The act, which was pushed through by an unusual coalition of envi-ronmentalists and sportfishermen large enough to dislodge the long-entrenched commercial-fishing lobby in Congress, shifted the burden of proof from scientists to fishermen; the equation had been inverted. After the SFA was passed in 1996, fish were to be assumed to be inherently scarce unless proven otherwise.

But what made the Sustainable Fisheries Act most significant is that for the first time since the era of industrial fishing began, it essentially *required* that overfishing be ended for every single American fish or shellfish. To this day neither the European Common Fisheries Policy nor the Canadian National Fisheries Policy has ever done such a thing. What the act said is that overfishing *is* a valid scientific concept. It does occur and has occurred, and it is our job to stop it. Indeed, for each individual stock of commercial fish that exists in American waters, the Sustainable Fisheries Act created specific goals and timelines for complete rebuilding of the population. It is now U.S. law that all commercial fish populations in the United States *must* be fully rebuilt by the year 2014.

The Sustainable Fisheries Act has actually changed things for the better, at least for fish. It has helped managers imagine the possibility of progress, particularly with gadiforms. When the act was passed, Georges Bank and Gulf of Maine cod stocks were at 12 percent of what fisheries scientists thought of as "rebuilt." Haddock, another gadiform, were even worse off. The result of the act and its unusual deadlines are impressive: it gave regulators the ability to impose the drastic measure of closing fishing grounds entirely should the rebuilding targets not be met on an annual basis. Ten years after the act's passage, Gulf of Maine codfish are now at 50 percent of their rebuilding goal and seem likely to achieve their goal by 2014. Gulf of Maine and Georges Bank haddock are now considered fully rebuilt. Unfortunately, Georges Bank cod, the stock I was fishing this past December, is not rebuilding with anywhere near that speed, and its target has been moved to 2026.

But while rebuilding targets are missed or postponed, it must be underlined again that the Sustainable Fisheries Act has, against immense pressure from fishing interests, allowed regulators to keep half of Georges Bank entirely closed to fishing. These closures have allowed the banks' ecosystem to stabilize. Trawlers no longer drag and redrag its fragile reef systems. Spawning cod no longer have to evade nets when they are at the most physiologically weak point in their life cycle. So even though the time horizon for rebuilding has been extended for Georges Bank cod to 2026, a year when my son will be twenty years old and I will be more of a burden than a helper on a fishing trip, there could still be a recovery. The stock has not slipped below a genetically defunct threshold; there exists the template for a refuge for fish in the long term. There is still hope.

With fish, though, hope must always be put in context. Around the time of the North American cod crises and the passage of the Sustainable Fisheries Act, Daniel Pauly, known as the most iconoclastic of leading contemporary marine biologists, coined the term "shifting baselines." When I came across the concept a little while back, I was struck by both its profound significance as well as its relative invisibility in the contemporary news cycle. Ghettoized within the insular realm of fisheries science, the theory has profound implications as a sociological phenomenon as much as a biological one.

The idea of shifting baselines is this: Every generation has its own, specific expectations of what "normal" is for nature, a baseline. One generation has one baseline for abundance while the next has a reduced version and the next reduced even more, and so on and so on until expectations of abundance are pathetically low. Before

Daniel Pauly expressed this generational memory loss as a scientific thesis, the fantastical catches of older fishermen could be written off as time-warped nostalgia. But Pauly has tabulated the historical catch data and shown that the good old days were in fact often much better. This is not nostalgia on the part of the old or lack of empathy on the part of the young. It is almost a willful forgetting—the means by which our species, generation by generation, finds reasonableness amid the irrational destruction of the greatest natural food system on earth.

My baseline, up until I started looking into codfish, was that codfish are fundamentally a fish that comes from far away, abundant on the slope of the continental shelf, a minimum of a two- to four-hour steam from land, and commercially pursued by distant offshore fleets. But as I started to look into codfish more closely, I was to come to realize that my baseline was considerably shifted from what nature had initially provided.

It turns out that codfish on Georges Bank and other offshore areas are populations of last resort—the head office of the cod operation with all its subsidiary franchises removed. And to a large extent the future of our codfish populations comes down to the question of whether humans can reconstruct a memory of the pattern of abundance and apply it to the future.

As Pauly's shifting baselines show, perceptions of abundance in human experience are relative. Even I doubted the existence of a cod shortage during my fishing trip to Georges Bank, because every time I dropped a jig to the bottom, a cod seemed to come up on my line. The modern marine conservationist must work against this limited perception and persuade fishermen that their immediate concept of abundance is a diminished one. Mark Kurlansky's *Cod* did this in a layman's way, creating a benchmark for a general readership that gives some evidence of the past abundance of cod. Science, though, requires more rigor and precision to quantify memory—a census, so to speak, not of the present but of the past. It was just such a census that a Maine fisherman named Ted Ames took up in 1999.

How do you tell an imbecile from a functional person?" Ted Ames asked me recently, his gentle Down East lilt making the word "person" come out as "puh-sin."

"When a functional person makes a mistake, he'll maybe try it once again, but after that he'll do it differently. The imbecile will do it over and over and over again."

Ames and I were talking about the management of cod in general, but we were discussing in particular the management of the cod that used to inhabit his native waters, the lovely rocky coast from Portland, Maine, on up past Boothbay Harbor to Stonington, all the way to the Canadian border. Ames is a former commercial cod fisherman, the son and grandson of cod fishermen, and it is his unique relationship to the *history* of cod that has allowed him to embark on a project of historical reconstruction that won him a MacArthur "genius" grant in 2005.

"After cod collapsed in 1995, I had already come ashore," Ames explained, using the term "coming ashore" in the Maine sense of giving up fishing. "The kids of the fishermen that I used to work with asked me if I would represent them in trying to get a fishery back along the coast. The government had just come up with a federal management plan for cod, and they'd come out saying there were historically only a couple hundred miles of codfish-spawning grounds on the coastal shelf. Fishermen throughout our area knew that was not the case. So I went around and interviewed fishermen and asked them where they caught ripe cod. We collectively found a thousand square miles, most of which was heretofore-unknown spawning grounds."

Over the last hundred years, one truth has come to emerge about fisheries management that few would dispute—you must know how many fish there were *before* fishing began to be able to predict how many fish there *can be* in the presence of fishing. If they have that historical data at their disposal, fisheries managers can get a sense of what they are striving to reclaim after fishing enters the system. Just as important, they can begin to recognize what represents a danger threshold below which a population should not be allowed to drop.

This model, however, has one basic, gargantuan flaw: it assumes that it is scientists who find fish first, make recommendations, and then model the methods of a would-be fishing fleet on those recommendations. In the entire history of fishing, this has probably never occurred. In real life it is fishermen, knowledgeable hunters who know their prey most fully, who find fish first. And when fishermen find a new stock of fish that has never been measured, they fish it and fish it damn hard before regulation can be put in place. The baseline disappears down the open end of a net trawl. As the scientists Boris Worm and Ransom Myers concluded in an oft-cited 2003 *Nature* paper on fish abundance, "Management schemes are usually implemented well after industrialized fishing has begun and only serve to stabilize fish biomass at low levels." In starker words, they tend to manage to preserve a status quo of scarcity, rather than to reestablish a historically correct abundance.

By conducting a series of interviews with seventy-, eighty-, and even ninety-year-old commercial fishermen whose early fishing days predate the advent of large-scale fishing technology, Ted Ames addressed this problem by establishing a different, more profound historical baseline. In his interviews he sought to identify extinct populations of cod. And what he has found through these interviews is that the population that is now called the Gulf of Maine stock is in fact the remnants of perhaps dozens of cod subpopulations that had at one time spawned up and down the Maine coast, often within sight of land. I'd always thought of cod as an offshore fish, pursuable only by a long trip in heavy seas. But what Ames found is that some cod even rushed up into the mouths of rivers, pursuing river herring in their runs from the open ocean.

Ames's research also highlighted another important element of the entire equation: it is not just overfishing that decimated cod. The destruction of cod's prey also played a crucial part. At one time runs of alewives, blueback, shad, and other fish in the herring family existed in rivers throughout coastal New England. Herring, like salmon, seek home rivers and must have unimpeded access to gravel beds in fresh water to spawn. But in the eighteenth and nineteenth centuries, low-level dams were erected all over the Northeast to power local textile mills. These days, in the tiny state of Connecticut alone, there are as many as five thousand dams (nobody actually knows the exact number). Even though mills are largely defunct and the dams no longer serve any practical purpose, they are still in place and herring populations are severely depressed. Cod cannot return to their former range in part because of a simple lack of food. Cod abundance, it turns out, is inherently linked to access to multiple sources of food over a long range. Take away the herring and you take away a key support beam for the cod kingdom. Cod's major food source becomes, by necessity, the prey on the distant coastal shelves; their food is plentiful only far out at sea, and therefore that is where cod survive best.

Ames concluded that "cod have complex population structures in the Northwest Atlantic with multiple subpopulations, and that managers have often failed to prevent the collapse and loss of spawning components in these heavily exploited fisheries." In other words, you must recognize that the relationship between large-scale offshore cod populations and small-scale coastal subpopulations is limited before you set rebuilding goals for the stock as a whole.

In light of these findings, Ames strongly asserts that the abandoned cod grounds that no longer support schools of codfish *must* once again become populated if the Gulf of Maine stock of cod is to be considered truly rebuilt and truly abundant. When I told him that fisheries managers I'd interviewed already consider Gulf of Maine cod 50 percent rebuilt, he laughed. "In that area many of the three thousand–odd full-time fishermen used to go fishing for cod and other groundfish during part of the year. Today there is one permit holder for cod, and he is the last active ground fisherman in a hundred and fifty miles of coastline. Is that a fishery that's fifty percent recovered?"

Nevertheless Ames believes that there are key, albeit tentative, biological signs of recovery occurring. This is why a new approach is necessary. Codfish *are* in fact creeping back to the old grounds; a few are showing up where before there were none at all. Ames believes that, managed correctly, within a decade the ancestral grounds could be recolonized. Perhaps they won't rebuild to the level that his grandfather would recognize as abundant, but we can hope at least for a stock that Ames would call "an awful lot of cod."

Thus, before we take the rebuilding goals of the Sustainable Fisheries Act as gospel, we must consider a larger historical picture and perhaps even look at it from the codfish's point of view. If you were a codfish, you would have dreams not of retaining a tiny redoubt of your former wealth but of reclaiming your entire kingdom, a kingdom that spans the area from landfall to the continental shelf in every temperate shore of the Northern Hemisphere.

How, then, to reclaim this kingdom? Certainly not through the present management approach, Ames argues, in which a distant fisheries council in New England makes what appear to be arbitrary and careless decisions about fishing grounds with which it is only superficially familiar. Ames sees the current plan to "rebuild" only the offshore codfish fishery and then turn over the permits to large industrial fishing companies as a model that has been tried again and again over the last thirty-five years and has always led inevitably from one collapse to another. No, what Ames thinks is the key to it all is allowing fishermen themselves a voice in managing the fish their livelihood depends on, giving a voice to the small-scale, diverse, artisanal fishermen who are deeply invested in the area and the fish they catch.

"The appropriate analogy is what the United States did with its system of government at the outset," Ames told me. "A system where federal, state, and county forms of government represent different levels of interest, all interacting with each other in a constructive way." This is a model that has been successfully deployed with Maine's other iconic animal: the lobster. There are now seven thousand Maine lobstermen, divided up into community units, each with intimate knowledge of and responsibility over a hyperlocal tract of ocean. Fishermen are not allowed to work an area unless they are residents, own a boat, and have evidenced a long-term commitment to their stock. And today lobster catches are booming. So much so that in 2008 the *New York Times* reported that seafood markets were experiencing a "lobster glut" and that prices were dropping by as much as four dollars a pound.

If such a system were put in place with codfish, Ames believes, the same kind of stewarded recovery could slowly take shape. But Ames also believes that simply turning out a new fleet of hyperlocal fishermen isn't enough. What he imagines is a new *breed* of fishermen, one that is as knowledgeable about the ecosystem as it is about fishing. "I think that management's biggest failing is that fishermen don't have the knowledge they need to manage a complex thing like a marine coastal ecosystem," Ames said, his voice rising in excitement. "Your right to fish shouldn't be based on whether or not you have a suitcase of money to buy a boat—it should be based on knowledge. Your right to fish should be won or lost on your willingness to comply with credible science. It would be great if the Fisheries Service said that no one could stay in a wheelhouse unless they pass a written or oral exam on fisheries ecology. I believe this can work. I believe that knowledge can change the dynamic."

As I listened to Ames explain all this, I realized that what he was imagining was something that has historically defined the most stable human/animal relationships. He was proposing that a stretch of territory be broken up into different lots, each with a fisherman possessing a deep, intimate knowledge of how many animals ranged around him, where they reproduced, how they were reared, how fast they grew. In such a system, a fisherman would adapt his fishing gear not just to catch more fish but to catch only the fish whose removal the ecosystem could support. And such a fisherman would stop fishing his territory not when he could no longer catch fish but rather when he observed clear biological signs that his swath of pasture needed to lie fallow.

I realized as I listened that he was proposing that fishermen, the world's last hunter-gatherers, become herders.

Fishermen as herders. I liked the idea. It seemed fair, organic, and balanced. It seemed like the kind of approach that would keep small-scale fishermen fishing and harbors working. Since fishermen are usually the people with the most intimate knowledge of fish behavior, maybe a little push in the direction of husbandry could fix the overfishing equation. Moreover, I could see the importance of having fishermen keep an anchor in the water. Humans have a way of overexploiting a resource if no one else is there to lay claim to it. An educated, thoughtful fisherman could be a guardian of sorts, someone who cherishes and understands the local ecology, someone who continues to make the ongoing argument against oil and gas exploration, or mining, or whatever else mankind will think up to exploit our coasts in the years to come.

And, slowly, the fisherman-as-steward concept appears to be getting a tenuous foothold in the market. The Cape Cod Commercial Hook Fishermen's Association, formed in 1991, has brought together a coterie of small-scale fishermen that use low-impact hook-and-line fishing gear to harvest cod and other groundfish from unclosed areas of Georges Bank. The Association des Ligneurs de la Pointe de Bretagne is doing similar things with the remaining wild European sea bass populations off the coast of France. In 2008 the association established a self-imposed fishing ban on sea bass from February 15 to March 15, the time when the fish is spawning and at its most vulnerable.

But there are some who would argue that the herding of fish is an intermediate step and that if we want to get truly efficient and productive, we have to turn the corner from herding and move on to outright cod farming, as we have done with salmon and sea bass—an entirely controlled path in which cod are taken out of the classic multi-input natural environment and put under monocultural human husbandry. When I began to investigate this, I found yet another instance where Mark Kurlansky's book *Cod* had a resonating impact. In a remote corner of the North Sea, a copy of Kurlansky's book had fallen into the hands of a Scotsman with the rather un-Scottish-sounding name of Karol Rzepkowski.

If you were to search for a place where the abundance of cod-fish could be reborn, the Shetland Islands would catch your eye on looks alone. A tiny archipelago stranded between Scotland and Norway, the Shetlands are considered by locals to be the United Kingdom's best-kept secret. The human population is scattered across a series of small settlements that seem drawn from the 1980s comedy *Local Hero*, where "crofters" still live in earthen huts and the treeless heath spreads out under the mild light of the midnight sun. Even the relatively populated capital, Lerwick, boasts clear green-blue waters with beautiful flowing strands of waving eelgrass.

The Shetlands have the highest per capita public investment in the United Kingdom. After oil and gas were discovered beneath near-shore waters in the 1970s, a remarkably disciplined town council forced British Petroleum to pay an uncharacteristically large portion of oil profits into the community. Given the islands' history of resistance, it is not surprising the locals won out.

Shetlanders are renowned for their resolve and independence, characteristics that date back to the time when the islands were discovered by two Viking lords. As legend has it, the two Norse noblemen had a bet between them—whoever touched the land of the islands first would get to claim them for his own. On realizing that his boat was falling behind as they approached landfall, the trailing nobleman took an ax from his belt, hacked off his hand, and threw it to shore before the other lord could touch dry land. Needless to say, the one-armed man won.

Nowadays Shetlanders fly the Norwegian flag nearly as often as they do the Union Jack, and while technically part of Scotland, a nation that historically bridles under the oppression of English power, Shetlanders often complain of the Scottish "invaders" who arrived in their islands and subjugated them just as the English later came and subjugated the Scots.

Karol Rzepkowski is not a native Shetlander. The child of émigrés who had fled Hitler, he grew up in Edinburgh, working in his father's Polish delicatessen. "It was a proper delicatessen," Rzepkowski told me when I went to visit him at the charming red barn that served as headquarters of Johnson Seafarms outside Lerwick. "There were big barrels of gherkins, properly salted." Following the demise of Europe's communist regimes in the late 1980s, Rzepkowski went to Eastern and Central Europe in the early 1990s and started an import-export business, where he traded in "everything from used clothes to oil pipelines." The business was a tremendous success, and after making a tidy profit Rzepkowski moved to the Caribbean, where he "semiretired" and taught scuba diving.

Rzepkowski is a frenetic person, and lolling on a Caribbean island did not suit his temperament. One day while sitting in his beach chair, he opened a book someone had given him that would catapult him out of retirement. The book was *Cod* by Mark Kurlansky. "*Cod* was the first time that I actually sat up and took notice just what an issue there was with cod in the wild today and how few there are."

Rzepkowski returned to Scotland, settling in the Shetlands, and thoughts of cod stayed in his head. In the early 2000s he answered an advertisement in a newspaper for Johnson Seafarms, a salmon aquaculture operation that needed a new manager. While he had no experience in fish farming, he had the idea to convert the operation over to cod farming, for he saw in cod an opportunity to grab the public's attention. "Cod is a new species to aquaculture," he told me. "The public is very aware of issues of the wild cod fisheries, so it's easier to actually feed the information about our product to the public. It has made our product much more high-profile."

Rzepkowski set about transforming Johnson Seafarms into a kind of living experiment on the validity of replacing a threatened wild species with an aquacultured version. And in order to show that these cod were as natural as possible, Rzepkowski decided to isolate all the things that had been criticized about industrial salmon farming and purge them from his process. Whereas Norwegians were trying to apply the model they had developed for salmon by investing millions of dollars in creating breeding lines to selectively create a highly productive race of cod, Rzepkowski refused to do any selective breeding at all. "I don't think it's a wise idea, especially when you're starting off with a new species and a new industry to launch headfirst into trying to create Super Cod. Why? Cod is super anyway. We don't need to turn it into Super Cod."

In addition to eschewing selective breeding, Rzepkowski chose to raise cod in accordance with organic standards that had been established by the Scottish Soil Association. Included within these standards was a ream of animal-welfare laws that required cod be granted the "five freedoms" as set out by Britain's Animal Welfare Council in 1992. These included freedom from overcrowding, freedom from hunger and thirst, freedom from disease, freedom to live life as close as they would have lived it had they been wild, and freedom from fear, distress, and mental suffering. Whereas some of Rzepkowski's competitors tried to dissuade cod from chewing their way out of their net pens by coating nets with unpleasant-tasting paints, Rzepkowski gave his cod chew toys.

But even with these strict standards being applied, the people at Johnson's still found themselves slaves to a hard-to-replicate wild system. In nature, cod tune their behavior to the dramatic changes in sunlight that occur over the course of the year at the high latitudes of the Northern Hemisphere. If it is the dark "halcyon days" of January that compel sea bass to spawn in the Mediterranean, in the North Sea it is the late-June solstice and the days of the *simmer dim* (the midnight sun in Shetland dialect) that cause a photosensitive organ in the cod's brain to trigger the release of gonadotropic hormone. This hormone makes cod stop growing and instead devote their energy to the growth of eggs (roe) and sperm (milt). The roe and milt develop throughout the fall and on into the depths of winter when the sun disappears from the sky entirely. Then, in January, as the first traces of spring sunshine penetrate the North Atlantic depths, the cod begin congregating in balls so tight that a single trawler, were it to locate them, could scoop up an entire school in a few drags. Soon the cod begin to form spawning columns, sometimes three hundred feet high, swarming around one another and choosing their mates.

This wild cod bacchanalia is an annual ritual that Europeans have come to rely upon. Today in most of Europe, two-thirds of all cod are eaten during the three-month spawning migration when the *skrei torsk,* or "wandering cod," the last truly healthy stock of wild cod in Europe, swim from the Barents Sea to the Norwegian fjords and the Scottish estuaries. This huge wild spawning migration presents a considerable problem for a would-be cod-farming entrepreneur. For if you are a cod farmer you must figure out a way to make your cod available when wild cod are scarce. And in order to do that, you must try to trick cod into spawning year-round so that there will always be a crop of cod reaching marketable size in any given month. At Johnson Seafarms this trick was achieved by artificially changing dawning of the midnight sun.

"Those fish over there are in the past, and those over there are in the future," a pleasant young man with a hoop earring who was managing Johnson's broodstock told me. In the breeding area of Johnson's, the cod had been separated into twelve different tanks, each one representing the light and photoperiod of a different month of the year. Since cod tune their spawning to the amount of daylight they receive each day, the staggering of light made it possible to have at least one group of cod spawning every month. Thus while wild cod might be available for only a few months out of the year, farmed cod could conceivably be available during all the months when wild cod can't be found.

Once cod do spawn in captivity, all the techniques of feed and juvenile rearing that were invented by the European sea bass farmers are applied in very similar fashion. Enriched rotifers are fed to hatchlings, enriched artemia to the next size up, and, finally, feed pellets derived from fish meal are fed to the larger fish until they reach market weight. At Johnson's, though, the standard fish meal of ground-up wild little fish had been replaced by "organic" fish meal approved by the Scottish Soil Association. In order to qualify as organic, fish meal must be derived from "off-cuts"—i.e, fins, bones, and other discarded parts of filleted commercially caught fish—the idea being that since these parts would have been discarded anyway, recycling them as fish meal extends the usefulness of the original wild-caught fish. But even though organic meal is in effect recycled fish, it is considerably more expensive than traditional fish feed, sometimes twice as expensive.

And so at Johnson's, with all its attention to animal welfare and politically correct feeds, the price of production was considerable. Hoping to achieve an economy of scale, Johnson's at the time of my visit was in the process of a massive ramp-up. The company had just built an airplane-hangar-size rearing facility in 2006, and when the juveniles raised in the airplane hangar were harvested in 2009, eight thousand tons of fish were anticipated—more than the current legal wild catch of Massachusetts's Georges Bank. In order to reclaim costs on this tremendous investment of time, money, and resources, Johnson's cod was set to sell at twenty dollars a pound, nearly double what wild-caught fish cost. But Rzepkowski presented the case as an environmental cause, not an economic one. "Aquaculture has the potential to apply the best practice with really no adverse effects to the environment," he told me shortly before I left the Shetlands. "You can actually mass-produce a product in an environmentally benign fashion if you just apply some rational thought to it."

Having the voice of wild-cod fishermen in one ear and cod aquaculturists in the other produced a certain dissonance, to say the least. Which was the right course for cod? A patient, carefully guided recovery with herder-fishermen gradually easing cod back to viability and onto the market? Or was such an approach an idealistic and unrealistic quest? Would the cycle of destruction that Mark Kurlansky showed to be so persistent in every cod stock in the world merely repeat itself if fishing continued? Would the world's increasing need to have more and more millions of pounds of whitefish every year on a regular basis simply prove too much of a temptation? Would it be better instead to give up on the profession of fishing altogether and use disciplined, organic practices to bring a husbanded product to market that could be predictably and sustainably raised?

Even with the strongest arguments for taming cod, I felt an instinctual allegiance to the wild fish and a kind of passive hostility to the farmed version. But I was at a loss as to how to express what it was that I found amiss with what taxonomists might someday call *Gadus domesticus*. I felt the need for an outside opinion. Someone who would sit down at the table with me, taste the different versions of the fish, and weigh with me the pluses and minuses of both. And for this reason I decided to conduct an experiment. I e-mailed Karol Rzepkowski and asked him to send me two farmed codfish from the Shetlands. At the same time, I put in a request to Whole Foods for several fillets of fresh "sustainably caught" wild codfish. I also asked a Norwegian nonorganic cod-farming operation to send me their samples, as a control group. And then, with a little trepidation, I picked up the phone and invited Mark Kurlansky to lunch.

Mark Kurlansky's opinion is frequently sought out. In the seafood realm, coastal towns and seafood restaurants ask him to taste their fish, while in the literary world authors and editors hunt him down to blurb the multitude of single-word-titled books that have proliferated in *Cod*'s wake. "I like to help other writers out," Kurlansky told me as I hurried to prepare the cod samples for our lunch, "but I don't really want to *read* all these books. I just got sent one on pigeons. I don't think I'm going to read the one on pigeons. I blurbed the one on rats. Do you realize how many of these books are coming out? There's *Mauve*, the color that changed the world, and . . . you know, it's just endless."

But unlike many of those authors, Kurlansky came by the subject of *Cod* honestly. Long before he wrote about fish, he caught them for a living. His university education, in fact, owes its existence largely to a commercial trawler out of Gloucester, Massachusetts, that employed him during his high-school and college summers, when he caught cod day in and day out off Georges Bank. Also, unlike many commercial fishermen, who tend to avoid eating seafood, his family always included codfish as a regular part of their weekly diet. So much so that cod wasn't even called cod at the dinner table. "We always called cod 'fish,' " Kurlansky told me. "If I asked my mother what she was cooking for dinner, if it was cod, she'd just say 'fish.' "

I let Kurlansky take a look at the Shetlands farmed codfish (I left one whole for this purpose). Opening my cooler, he reached in and grabbed the fish by its gill. He fingered the barbel under its chin—the tastebud-rich flap of skin that cod and other gadiforms use as an external tongue to taste the bottom as they swim along. Regarding the cod with some suspicion, he pulled on the barbel and looked it in the eye.

"What are you doing?" I asked.

"Just trying to see if it looks like a real cod."

As Kurlansky continued to handle the cod, I went over and plated the fish that I had finished baking. Arranging them on the table in an arc so that I could remember which was which, I called Kurlansky to the table. He flaked them apart and smelled them as one would a fine wine. Putting down his fork, he made his assessments.

"Number one [the wild fish] had a good, loose flake. Like a textbook-perfect flake. Nice, but no taste. The second one [Norwegian conventionally farmed] had a very strong flavor. The flake was not good. There was a slight metallic taste, but it was not unpleasant. The third [Karol Rzepkowski's Shetland organic-farmed] I found to have a very strong, unpleasant flavor. I'm not sure of what. It had an okay flake, not as okay as the first."

Before revealing which was which, I asked him to pick a winner. "I might go with fish number one for the texture and the second for the flavor." I pressed him to pick just one. "I might go with the second one," he said finally. "Then again," Kurlansky joked, throwing up his hands, "I might pick fish number four."

When I at last revealed which fish was which, he gravitated back to the qualities of fish number one, the wild fish.

"The first one was exactly like a cod is supposed to be. I was just sort of amazed that it had no taste. But I guess cod is like that sometimes."

The more he thought about it, the more he became focused on the texture.

"If you talk to people in cod cultures, what they talk about is the texture, not the flavor. I just find it so interesting that the texture was the thing that was off with the farmed fish. In hindsight, I think of course it's about muscle. Those farmed fish aren't living a cod's life."

The Kurlansky taste test, over and above the scoring, pointed to something that I had found disquieting when I'd visited Karol Rzepkowski's cod-farming operation. Listening to Rzepkowski talk about rapacious overfishing, catch quotas, and the importance of an organic approach, I couldn't help but notice how, though he was clearly very well intentioned, those intentions were being warped by the need to produce a sustainable business model. It was only after I bade good-bye to Kurlansky and reflected on the whole of my farmed- and wild-cod experience that I started to understand what it was that I found bothersome.

With all fish facing declines, scarcity can have a weird effect—it can become a kind of marketing tool. Even though cod should be so common that a person like Mrs. Kurlansky should call it simply "fish," Rzepkowski saw the public perception of cod-in-trouble as a way of raising the animal's profile, of making it somehow special and different, "COD" rather than just familiar old "fish." Given the historical, biological, and economic roles that cod has played, I wasn't quite sure that was the right role for the animal. Cod was most distinctly *not* a king, not a holiday fish; it was an everyday fish. And its lack of extreme abundance in the wild was something we needed to address and fix, not something we could merely replace with a farmed product.

And in the end, even though Rzepkowski had in his mind created a more sustainable model, his farmed codfish proved to be even more fragile than the wild fish of Georges Bank. The year after my visit to the Shetlands, Johnson Seafarms cod was rebranded as "No Catch," and a bizarre series of advertisements followed. The Web site was particularly weird—on the site, Rzepkowski and his colleagues appeared as video faces set atop strange, Monty Pythonesque moving stick figures. Overfishing is cheekily reduced to a quick cartoon of a boat swooping in, grabbing a school of cod, and leaving a handful in return. The cod chew toys are shown in garish colors, and animated cod leap out of the water in great numbers with every click of the mouse.

But all the hype of the Johnson's–turned–No Catch cod could not overcome one crucial problem with trying to farm cod commercially. Cod have a bony, oversize head that accommodates a giant, oversize mouth: a head and mouth designed for conditions where prey is even more abundant than the already very abundant predator. Cod evolved this morphology in order to swim slowly with mouth agape, always ready to vacuum in everything from lobsters to herring so long as not too much energy is expended. But the average Western consumer wants nothing to do with a cod's head. Huge amounts of money in a cod-farming operation are thus uselessly invested in growing thousands of tons of cod heads that no one wants. This, combined with the fact that it takes an exceptionally long time in aquaculture terms—three years, as opposed to two or less with salmon or sea bass—to grow a mature cod, adds to the difficulties. With cod you just don't get enough meat fast enough for your money.

Lastly—and this is the argument that conservation organizations like the Pew Charitable Trusts and the Monterey Bay Aquarium lodge against the taming of any large predatory fish, be it European sea bass or salmon—the final crop of farmed cod produces a net marine protein loss for the sea. No Catch may have been using by-products from fish processing for its animal feed, but those discards could be more effectively used in faster-growing, more efficient animals like barramundi. It might be too late to rid the world of carnivorous, feed-intensive, farmed salmon (the industry is too large, too dug in to simply end), but that does not mean we should necessarily bring yet another carnivorous, feed-intensive fish into domestication. In a world where marine protein is getting scarce, playing around with a species merely because that species is well known and well liked can no longer really be justified. The outputs and the inputs need to be considered with equal weight.

Things at No Catch started going downhill in 2008. The huge capital (mostly from all those petrodollars sloshing around the Shetlands community) slowed to a trickle, and the money needed to keep the airport hangar filled with young cod became too much.

Most aquaculture operations, especially those trying out new species, begin their lives as losing ventures, and then, slowly, some begin to make a little money. Johnson's cod farm just couldn't make up the price. By the winter of early 2008, the company was in receivership. There was hope for a while that somehow it would continue operation under different management. But even that fell apart beneath the weight of tens of millions of pounds sterling of debt. Soon the company was broken up and the codfish were slaughtered and sold at cost. The remainder of No Catch's resources was divvied up among its competitors, with the majority going to Norway, where a more industrial form of codfish farming was trying to make its way to market.

The Norwegians may yet succeed with farmed cod. Production has increased yearly, and in Norway a point of price parity has been reached with wild and farmed versions of the species. But even though Norwegian cod are cheaper (the price of, say, sirloin versus No Catch's filet mignon), what the world needs is something several orders of magnitude less expensive—the seafood equivalent of ground chuck. The world needs Mrs. Kurlansky's "fish," not "COD," and people want to get "fish" at the prices they believe "fish" is worth. And for this the public has discovered that it has had to look for something else entirely.

F inding a replacement for that classic "flake" of codfish that Mark Kurlansky so appreciated, something that *did* "live a cod's life" and had the "mouth feel" that fish eaters would find familiar and that furthermore would freeze and travel well was not an easy task. Not only would the flake have to be right, but, more important, there would have to be a lot of that flake to go around. With farmed cod failing to meet either of those requirements and with wild cod now hemmed in by the Sustainable Fisheries Act, the industry, in the wake of the Georges Bank and Grand Banks collapses, would have to find another species altogether. But whatever species they lit upon, the big supermarket chains would have the problem of relentless regular demand to contend with.

The modern supermarket has a basic internal ecology that finds an equilibrium between what its suppliers can produce and what its customers can consume. Terrestrial food, for the most part, fits well into that equation. Purchase orders for meat and bread can be put in months in advance with the full confidence that the hoofed mammals and monocot plants that we grow to produce those commodities will be present in sufficient numbers when it comes time for orders to be completed. The essential inputs are known (feed, fertilizer, water, land), the risks (disease, drought, heat, cold) are increasingly calculable and addressable, and the outputs (pounds of feed needed to produce a pound of edible beef) are measurable down to a decimal place.

But cod and other wild fish are something else. The industrial food sector must work around the vagaries of a natural system. Any number of factors, ranging from an overactive oceanic gyre to a ripple in the population of herring prey fish, may throw a wild fishery at least temporarily out of whack. And so in the global wild whitefish market, there are, in effect, two systems running side by side: the human-focused, need-driven system where demand remains constant; and the diverse, disparate natural marine system that varies from year to year as a result of a plethora of uncontrollable variables.

Trying to find a sensible place for industry in this situation requires a population of fish, somewhere in the world, so abundant that a massive, consistent deduction will not cause an implosion of the stock—an implosion that had already occurred on the Grand Banks and Georges Bank. By the late 1990s, when large retailers were looking for a replacement for cod, they were increasingly facing pressure from the environmental community not to repeat the same dynamic that had ruined the cod fisheries. A replacement fish would have to be found that at least had the appearance of sustainability, as determined by some objective source. This need was particularly high for the world's largest seafood buyer, the English-Dutch corporation Unilever.

Unilever came into being when Lever Brothers, a British soap manufacturer, merged with the Dutch margarine producer Margarine Unie in 1930. Over the years Unilever developed from a retailer into a brand consolidator, and in 1995 it purchased what was perhaps the United States' best-known seafood brand, Gorton's of Gloucester. The purchase was made, however, just after the Georges Bank codfish fishery had been closed, and Unilever immediately found itself in the hornet's nest of a rising ocean-conservation movement. Reacting to fisheries crises in both the United States and Europe, Greenpeace began organizing a campaign against Unilever, threatening a boycott of its seafood products.

But Unilever managed to pull off one of the greatest reversals in the history of the modern-day green movement. Applying market principles to the nonprofit world, it sought out a partnership with another global environmental charity, the World Wildlife Fund, and jointly fashioned a new nonprofit called the Marine Stewardship Council (MSC), devoted specifically to the task of identifying sustainable stocks of fish around the world and setting standards for fishing those stocks.

At first MSC limited itself to certifying small fisheries. And indeed this has been a very good and positive thing for small-scale fishing communities and the stocks they fish. But large retailers like Unilever needed a much more populous whitefish. An "industrial" fish that could be caught in large numbers but which could also qualify as being sustainable. It was this need that led them to New Zealand and an ignominious fish called the hoki.

The hoki is a gadiform descended from a fish that ended up in the Southern Hemisphere after the great gadiform radiation tens of millions of years ago. It is a cod-size, silvery-skinned, white-fleshed fish that was as abundant as the Georges Bank cod once were. It had been left relatively untouched until New Zealand fishermen pioneered technology for deepwater-fisheries extraction. It seemed to be exactly what sustainable fish marketers were looking for.

But right from the start, in 2001, the sustainability-certification process for the hoki drew fire. MSC does not directly certify fisheries; rather, the applicant fishery contracts the certification to a third-party certifying body. In the case of the hoki it was a coalition of fishing companies united under the name the Hoki Fishery Management Council that contracted out the certification to the Netherlands-based consulting firm SGS Product and Process Certification (SGS). In the MSC process, the third-party certifier evaluates the fish on a range of different criteria under three main categories: the sustainability of the target fish stock; the environmental impacts of the fishing (including accidental bycatch of seabirds, marine mammals, and other fish and the impact of the fishing techniques on the ocean environment); and finally how well the fishery is overseen and managed. Collectively, the scores of an MSC certification have to add up to eighty points out of a hundred for each of the three principals for a fishery to pass. Those who witnessed the process with the hoki felt that the process was lacking.

"We were involved in the certification process, and we didn't believe the fishery should be certified," said Kevin Hackwell, the advocacy manager of a well-established New Zealand conservation organization called the Royal Forest and Bird Protection Society. "We objected to the scoring that it got from the certification body. We took it to appeal. This was the first time an appeal had ever happened in an MSC certification process. And we lost that appeal on the basis that while the Objections Panel agreed at the time that the assessment of the fishery did not meet the MSC standard and therefore shouldn't have been certified, nevertheless it considered by the time the objection had been heard things had changed enough to let it through. We were very frustrated as this decision did not match the MSC's stated process. Our argument that it had been wrongly certified had been upheld in the objections process, but the fishery nevertheless got certified. The process was a farce." In an internal e-mail to the head of World Wildlife Fund International, the WWF New Zealand chief executive Jo Breese echoed this sentiment. "At this stage it appears likely that we will not be able to support the cerification process and outcomes," Breese wrote. "If we are asked by the media we will be forced to publicly criticize the process and possibly the outcomes."

Shortly thereafter hoki received MSC certification. "All the way through, we were saying that the allowable take was too high," Hackwell continued, "and sure enough, we were right." The fishery collapsed and the annual allowable quota dropped from 250,000 tons at the time of the MSC certification to 90,000 tons just a few years later. The hoki was decertified.

In 2005 the hoki fishery began the process of recertification. And once again Forest and Bird found the process troubling. According to Hackwell, in the first round, the hoki was scored by the certification body to be just at the eighty-point certification threshold on two of the three principals of the MSC's criteria. "But," Hackwell continued, "after detailed submissions from Forest and Bird and World Wildlife Fund's New Zealand chapter, several of the indicators had their scores reduced. This reduction would have brought the fishery below the eighty-point threshold. But the certification body lifted the score for several other criteria that had not even been the subject of comment previously. There was no reason given for increasing these scores and coincidentally these increases were just enough to push the fishery past the threshold by two one-thousandths of a point. Both Forest and Bird and WWF New Zealand took an objection against the recertification and in a weird echo of the 2001 objection the Objections Panel acknowledged that they considered the fishery should not have been recertified, but refrained from ruling to this effect." The hoki fishery was recertified as sustainable and retains its certification to this day.

Across the world, in another hemisphere altogether, the fishery of another abundant community of gadiforms certified by MSC is drawing criticism from environmental organizations, particularly Greenpeace. Alaska pollock are today the largest source of wild whitefish in the world. Nearly 2 billion pounds of the fish came to market in 2009. If you have eaten a fish stick, a Filet-O-Fish sandwich, a California roll, or any other processed white fish, you have eaten Alaska pollock. And, increasingly in Europe, Alaska pollock are being sold as flash-frozen whole fillets, a niche that had once historically been nearly the exclusive domain of cod. The fishery was first certified by MSC in 2005.

But even the huge numbers of Alaska pollock can at times show vulnerability. This year the stock assessment recommended that the harvest be cut in half, and, like the hoki industry before it, the pollock industry has been drawn into a period of recertification with MSC. The At-Sea Processors Association that speaks for the Alaska pollock fishery asserts that these are natural fluctuations in a natural system. This may very well be the case; even in healthy gadiform populations, fluctuations in population density can vary as much as 50 percent year to year. When I mentioned to At-Sea's director of public relations, Jim Gilmore, that Greenpeace had found the industry's sustainability qualifications questionable, Gilmore told me, "Greenpeace does not acknowledge that environmental conditions, particularly water temperatures, have a much greater impact on pollock population size than pollock harvests. Nor does Greenpeace note other sources of pollock mortality. One of my favorite 'gee whiz' facts in the November 2009 pollock stock-assessment document is that adult pollock are estimated to consume more than two and a half million metric tons of small pollock, or three times more than the 2009 harvest level." Furthermore, Gilmore asserts, Alaska, unlike New England, has a long history of limiting the number of vessels that can enter the fishery and has historically maintained large areas closed to fishing.

All this is true, and the Alaska pollock industry may indeed be worthy of its sustainable MSC rating. But when listening to the assertions of a major seafood purveyor, it is always important to remember the other "ecology" at play in fisheries—that of the global supermarket. An ecology that must have a constant supply of fish to keep functioning, no matter what natural limitations dictate. People who witnessed the cod collapses of the 1990s see many echoes of the past in the pollock industry. When I asked Ted Ames, the former cod fisherman from Maine who advocates small-scale, artisanal herder-fishermen, what he thought about the behavior of the large companies that hold all the large permits for the pollock fishery in Alaska, he chortled. "An old friend named Fulton Gross summed up this kind of thing in a pretty neat way. 'Remember one thing,' he told me. 'Never get between a fat hog and a trough. He'll run you over every time.'"

Greenpeace, the original campaigner against Unilever, is continuing its pollock and hoki campaigns and has said current allowable pollock catches, already cut nearly in half, should be reduced even further. But just as with the New England cod fishery, the concentration of power into a few dominant hands makes the industry a muscular opponent. Today only two companies, Trident Seafoods and Icicle Seafoods, account for virtually all the Bering Sea pollock inshore processing, and after twenty years of consolidation just five companies own all the fishery's vessels. As Geoff Shester, the senior science manager of perhaps the most influential list of sustainable seafood, Monterey Bay Aquarium's Seafood Watch, told me, "The Alaska pollock industry is just a huge player. Because there is so much money at stake, they have enough political influence to seek exemptions from regulations to protect ecosystems that might otherwise be costly to industry." In 2006 the pollock fishery was exempted from "Essential Fish Habitat" fishing closures that were instituted to protect the seafloor habitat. The pollock industry puts forth that it is primarily a "mid water trawl" fishery, catching fish far above the sea floor and doing little damage to sensitive ecosystems at the ocean's bottom. But Shester disagrees. "The U.S. National Marine Fisheries Service estimated they [the pollock industry] are operating on the seafloor 44 percent of the time and causing a greater overall impact to the Bering Sea seafloor than all other bottom trawling combined. Furthermore," Shester continued, "at a time when pollock stocks are at their lowest level in thirty years, the North Pacific Fishery Management Council in December of 2009 decided to set the most aggressive quotas allowed by law."

So can we call the large-scale industrial fishing of pollock a replacement for the already overfished cod? Maybe, or maybe not. Pollock are fast to reproduce and endemically abundant. But an annual harvest of 2 billion pounds of fish is a lot of wildlife to remove from an ecosystem every year. The Monterey Bay Aquarium has downgraded pollock from "best choice" to "good alternative" in their global seafood ratings card. Monterey Bay Aquarium's Shester maintains, "We continue to recommend pollock as a sustainable choice to consumers and businesses," and, relative to other whitefish grounds in the world, Alaska pollock is a well-managed fishery. But Shester's substantial concerns are noteworthy. Greenpeace is more forthcoming, calling for much more drastic cuts in pollock catches. And because no food purveyor wants to be associated with a Greenpeace campaign or the damaging of an ecosystem, there are some murmurs within the food industry about what to do next. Today several major restaurant chains and supermarkets are beginning to look beyond the gadiforms pollock and hoki—back to where both fishing and aquaculture began: fresh water.

Not long after my taste test with Mark Kurlansky, I found myself in the company of a Mr. Vo Thanh Khon in the southern Vietnamese city of Can Tho. On a blistering day in May, Mr. Khon led me through the sparkling new industrial park owned by the aquaculture company Bianfishco. Passing by a neatly manicured landscape of pruned palm trees, lower plantings of many varieties, and little magenta nine-o'clock flowers blooming brightly in the interstices, Khon, a short, dapper man in a pressed white shirt and black slacks, stretched out his hand and encouraged me to appreciate my surroundings.

"You are looking at our virtue," he told me.

We speedboated quickly across the Mekong, the largest river in Southeast Asia, and Khon and I walked over to a perfectly square pond a few yards inland from the riverbank. He gave a signal to a man in a reed hat piloting a small skiff, who then began shoveling yellow, dime-size pellets out the back of the boat. A few dimples appeared on the surface of the water. There was a splash here and there. Then, as if the entire pond were moving to engulf the skiff, the water erupted into a roaring froth, drenching the boatman and even sprinkling those of us standing twenty yards away on shore. Looking into the water now was like watching an M. C. Escher lithograph come to life. The water was boiling with two-foot-long fish, gray on top, white on the bottom, with faces that recalled a sentient but slightly dim-witted minor character in a *Star Wars* sequel, creatures that interlocked and overlapped and wriggled every which way. As I appraised them, along with the brochure of Mr. Khon's Bianfishco, the corporate motto "Pangasius is our nature!" struck me as more than a little bit ironic. Mr. Khon, however, is not an ironic man, and he smiled broadly at the roaring sound of the feeding frenzy. It was almost literally the sound of money earning interest.

The fish that were in Mr. Khon's pond are known internationally by their genus name *Pangasius* and locally as tra. If records from Vietnamese growers and government officials are to be believed, tra may be the most productive food fish on earth. Whereas an acre of codfish net pens will produce about ten thousand pounds of cod in a good year, that same acre in Vietnam will churn out half a million pounds of tra. This incredible tendency toward abundance has made the fish into the fourth most common aquaculture product in the world. From 50 million pounds in 1997, annual production has grown to well over 2.2 *billion* pounds, a large portion of which goes to Europe. Production is still growing, and no one can quite say where the upper limit will be.

Even so, many suspicions were raised when tra first entered the European marketplace not long after the cod crises reached their peak in the late 1990s. And much of that suspicion emanates from how they were found and why they were first farmed. To illustrate, it's worth repeating a joke my translator told me while we were motoring across the Mekong.

Question: "How do you tell a farmed fish from a wild fish?"

Answer: "The farmed fish is cross-eyed from staring up at the hole in the outhouse."

The *Pangasius* genus, as well as quite a bit of freshwater aquaculture, can in fact trace its relationship with humans back to the privy. Tra were first introduced several hundred years ago into "latrine ponds"—stagnant bodies of water that peasant families maintained adjacent to the Mekong. *Pangasius* in these ponds fed on . . . well, let's call it "decaying organic matter." In addition to managing human waste, an added advantage of these fish was that when they were large enough, they could be sold to one of the many floating markets that line the banks of the Mekong all the way up to the Cambodian border.

This was the state of things for many years while Southeast Asia remained war-torn, isolated, and dependent upon subsistence food production. But beginning in the 1970s, ethnic Vietnamese living on Lake Tonlé Sap in Cambodia began intensively culturing fish in floating cages underneath their houseboats. They tried many different fish at first but through a process of elimination gradually arrived at *Pangasius*.

"When farmers first collected samples for farming," Dr. Nguyen Thanh Phuong, dean of the College of Aquaculture and Fisheries at Can Tho University, told me, "they would collect juvenile fish randomly. They didn't even know what kind of fish they had in their jars. But when the oxygen ran out in the jars, all the fingerlings died. Except the *Pangasius*."

There were two species of *Pangasius* that survived the jar test— *Pangasius bocourti*, known locally as basa (meaning "three balls," because when cut lengthwise the flesh has three globes of fat distributed evenly around the spinal column), and *Pangasius hypophthalmus*, or tra. Initially it was basa that looked as though it would become the fish of choice. It adapted well to a variety of cooking methods and had a higher fat content than tra, which Southeast Asians value more than a dryer, flakier consistency.

But when ethnic Vietnamese began returning to Vietnam from Cambodia after the war between the two countries subsided, they started refining their aquaculture techniques. By the 1990s they gradually realized that tra took better to conditions in Vietnam. Unlike the lake environment in Cambodia, the Mekong Basin in Vietnam floods intensely every year. The entire flow of the Mekong is replaced during flood season, and stranded ponds are created adjacent to the main channel of the river. In these ponds it was found that conditions for raising fish could be better controlled than in the main stem of the Mekong. Pond culture didn't have the disease problems and the water pH shifts that occasionally killed off fish in the open river. But the more desirable basa needed flowing water to prosper and didn't do well in the ponds. Tra, on the contrary, seemed to thrive in stagnation. And while basa would die if too many fish were stocked in a pond, tra didn't mind close quarters at all. The only thing tra would do differently in these high-density environments when oxygen levels in the water dwindled was to occasionally rise to the surface and stick their alien-looking mouths out of the water.

Tra, it turned out, can breathe air.

The idea of using a farmed freshwater fish to substitute for a wild, oceanic gadiform has been something that seafood marketers had been considering for some time. And in Europe, where cod is used in a fairly diverse range of cookery, slipping a slightly different fish into the culinary slot was possible. Sautéed or baked, tra can for not-so-discerning palates seem like cod or at least like Mrs. Kurlansky's "fish." But in the more classic American use of cod, battered and deep-fried, tra lack the "mouth feel" of cod. The flesh is slightly oilier, slightly more substantial, more like a bass than a cod. In fact, in the jumbled-up world of seafood import and export, Vietnamese tra occasionally gets labeled as that other Vietnamese catfish species, basa, and is then slotted into the culinary niche where bass should be. Back in Greece, the sea bass farmer Thanasis Frentzos lamented to me one evening that the Vietnamese could potentially cause the death of the Greek sea bass industry. "Sometimes they mean to write 'basa' on a crate of Vietnamese fish, and then someone decides to replace the *a* at the end with another *s*, and then you have 'bass.'"

But tra are not the only superproductive freshwater fish out there. One class of fish in particular has arisen that is capable of assuming cod's role as an industrial fish both in texture and in quantity. Whereas the tra's key to abundance is tolerance for ultra-high stocking densities, another fish, called the tilapia, has made a name for itself in the abundance arena through its reproductive strategies.

Unlike cod and pollock, which hurl millions of small eggs far and wide ("broadcast spawners," in fisheries-science parlance), tilapia are of a family of fish called cichlids who have a tendency to be "mouth spawners." They lay fewer eggs than cod, but females typically gather up those eggs, once fertilized, in their mouths and protect them until they have passed the fragile early-larval stage. As a result the average tilapia produces many more adults in the end than the average cod, and tilapia have an ability to greatly multiply their numbers exceedingly fast. It is a reproductive strategy suited much better to a twenty-first-century human-dominated world than is the random "trust in nature" approach utilized by cod and other gadiforms.

Tilapia, like tra, saw a gradual buildup in abundance in the second half of the twentieth century, but, as with tra, its initial expansion occurred primarily in the developing world. Most tilapia hail from the Nile but were first spread beyond Africa when the Japanese army blockaded Indonesia during World War II. At the time, Indonesian fish farmers relied on a fish called milkfish for their aquaculture farms, but with the blockade they couldn't access the milkfish broodstock, which became stranded behind enemy lines. American forces were able to get a few stray tilapia to the Indonesians, and they soon found that tilapia grew nearly twice as fast as milkfish.

After the war, when the Peace Corps was born and the United States Agency for International Development implemented hunger-relief programs in postcolonial countries around the world, tilapia were seen as a solution to the world protein deficit. Not only did they reproduce with great abandon and without any help from humans, they technically required no food whatsoever. Tilapia, like tra in their native state, are filter feeders, able to live solely off elements of human waste, algae, and other microscopic plankton. So with tilapia, poor farmers, whose only resource besides land might be a stagnant patch of muddy water, suddenly had the chance to add protein to their diet with very little effort.

Early Peace Corps volunteers became unabashed tilapia enthusiasts. And when they left the Peace Corps and moved into the for-profit world, they saw an opportunity to turn the fish into a moneymaker. "It was like this miracle fish," a former Peace Corps worker–turned–tilapia entrepreneur named Mike Picchietti told me recently. "We thought we could make this fish into a major business. But it ended up taking a long, long time."

Until the 1990s both tra and tilapia remained "development" fish—Third World menu items that would have no real market impact in Europe or the United States. Partly this was because the fish had no brand identity in First World nations. (One aquaculturist told me that when he first heard the word "tilapia," he thought it was a stomach disease.) But another major factor had to do with a phenomenon that plagues all freshwater-fish farmers, something known as "off-flavor." Off-flavor occurs in stagnant fresh water when certain varieties of blue-green algae bloom and emit a compound called geosim, from the Greek *geo*, meaning "earth." Fish inhabiting these blooms temporarily take on the taste of geosim, a harmless but earthy flavor that most diners find unpleasantly muddy. In fact, off-flavor is one of the key reasons that many consumers have stayed resistant to farmed fish. Any fish can develop off-flavor, though freshwater species are more susceptible to it. Mastering the problem turns out to be the key to producing a widely acceptable product.

In the 1990s both tilapia and tra went through a revolution that transformed them from Third World to First World table fare. By 1994, tilapia, with their incredible fertility, had spread to Latin America. In a few instances, they were associated with the cocaine trade. South American Indians farming coca leaves in Colombia also came to farm tilapia. And while the coca crop was of paramount importance to drug lords operating in the region, tilapia could also serve their purposes. With millions of dollars of excess cash that needed to be laundered, they saw in tilapia an opportunity both to improve the lives of their growers and also to cleanse their drug take. There was an added benefit to coca growers. As one fish farmer told me, asking not to be identified by name for obvious reasons, "If you put a Gel-Pak of cocaine in a crate full of tilapia fillets, can a drug-sniffing dog find it? Nope."

What both tilapia farmers in Latin America and tra farmers and Vietnam realized was that if they managed to secure a constant supply of clear, flowing water, devoid of algal blooms, and if they fed their fish a diet of corn and soy instead of letting them subsist on waste and algae, they were able to control the off-flavor in their product. Instead of tasting like mud, tilapia and tra, by the late 1990s, tasted like nothing.

"The thing people like about tilapia," Picchietti told me, "is that it doesn't taste like fish. We often like to say it's the unfishy fish." Which, if the example of cod is considered, might ultimately be what the world is looking for. In twenty years tilapia production has tripled from 2 billion to nearly 6 billion pounds annually and is anticipated to grow another 10 percent in the next year alone. It is a naturally abundant, flavor-neutral product that is versatile in the kitchen and easy on the wallet. Indeed, these very qualities (or lack of qualities) have now made tilapia scale the highest heights of cheap food. After developing a patent-pending marinade that gives tilapia a "taste like pollock," HQ Sustainable Maritime Industries, a tilapia grower whose very robust production comes primarily from China, has concluded successful negotiations with one of the largest fast-food chains in the world. Although I cannot name the chain in these pages, suffice it to say it is likely that all of us will have the option of eating a fast-food tilapia sandwich in the not-too-distant future.

The Aquaculture Dialogues is a series of eight working groups convened by the World Wildlife Fund with the goal of creating standards for the fish-farming industry. In 2008 I sat in on a session of the Tilapia Dialogue in Washington, D.C. Early on in the discussion a motion was put before the gathered assembly of tilapia farmers, nonprofit organizations, various and sundry scientists and onlookers like myself that from here forward, people in the tilapia industry should "seek to prevent the global spread of tilapia beyond its already established range." A chuckle rippled around the room. "Too late!" a farmer from Pennsylvania said with a laugh. "Already happened," said another.

It is one of the great ironies of the modern-day seafood world that humanity is desperately trying to figure out a way to boost the numbers of one fish, cod, and, as an indirect result of the cod shortages, we are doing everything we can to keep another fish, tilapia, from multiplying and spreading too quickly. Tilapia is overstepping its ecological bounds in nearly every corner of the globe and is considered a most invasive species. The fish live mostly in fresh water, and freshwater bodies around the world are increasingly dominated by them. Tilapia can also adapt to more-saline conditions than most freshwater fish can tolerate, and as a result often find their way into brackish waters near river mouths.

There is now an active effort to try to put the biological tilapia genie back in the bottle. In Australia a $16 million campaign has been launched to keep tilapia out of the island continent. There are considerable obstacles. Chinese immigrants consider it good luck before sitting down to a meal of fish to release a live fish into the water. Because they are so hardy and require so little oxygen, tilapia are perfectly suited to take advantage of this cultural tradition. They are often delivered live to Asian markets throughout the United States, Europe, and Oceania, and when they arrive, it is not uncommon for Chinese immigrants to drop one or two into a local lake or stream. From there it is just a matter of time before they proliferate at frightening speed and reach a biological maximum.

In the United States and Europe, the ultimate range of tilapia has been restricted by a climatological zone. Tilapia die if water temperatures fall below fifty degrees Fahrenheit for more than a month, and so even when they are purposefully grown north of the Deep South in the summer, they die off when winter sets in. But winters are getting milder and shorter, and with each successively warm year tilapia inch a little bit farther north.

Meanwhile in the ocean, climate change is causing shifts among all the gadiforms. The patterns of ocean currents and weather in general are slowly changing, and the gadiforms that we once depended on for our flavor-neutral white-flesh fish are drifting away from us. The huge schools of Alaska pollock, whether they are fished sustainably or not, are undergoing substantial changes, and there are some indications that they are moving out of the territorial waters of the United States and drifting over to the considerably less regulated coast of Russia.

So we find ourselves, then, at the crossroads of change with all this "whitefish"—African tilapia spreading pell-mell around the world, Vietnamese tra inserting themselves into all kinds of markets for all kinds of fish, Alaska pollock and New Zealand hoki being presented to us as a "good" industrial wild fish but nevertheless declining alarmingly. And of course the fragile stocks of New England, Canadian, and European codfish hanging on to viability by the very tips of their barbels.

If we are to choose one fish from these many candidate species to serve as the backbone of our whitefish needs, which one shall it be? Do we just take all of them and not worry about parsing them, put all of them down the tube marked "flaky white" and hope for the best?

Looking across the different orders, families, genera, and ecosystems that these fish represent, at the risk of sounding shamanistic, it seems to me that nature is telling us something important about these fish and how we should use them. What it seems to be saying is that wild oceanic fish like cod and soon maybe pollock and hoki have vulnerabilities that make the industrial use of their flesh problematic. That indeed the very notion of "industrial fishing" may have to be reexamined. Perhaps in an earlier era, when the world population was a fifth of what it is today, when fishing gear was smaller and less intrusive, and when river and estuary systems produced a bounty of forage fish like herring, perhaps then a multinational industry could sit upon the vagaries of constantly fluctuating wild populations of cod, pollock, and hoki.

But contemporary demand is so large that any natural system is going to be taxed when subject to humanity's global appetite. The historical fragility of objective science in the face of the need to show corporate profits builds inherent conflicts into the system. With wild fisheries, when profit motives get too strong, even the soundest of scientific management schemes have a tendency to bend under the weight of demand. Scientific bases, however, do not work if they are bent. If they are forced to bend, fish populations inevitably enter into a spiral of decline that can lead to genetic collapse and irreparable damage to a hugely important food system. If there is even the slightest chance that fishing could cause the collapse of a stock, it goes against humanity's hope for long-term survival to continue the practice of industrial fishing.

At the same time, nature seems be telling us, in the form of tra and tilapia, that we might have found an industrial fish that works. These fish can live side by side in our environment. They grow fast. They are hugely adaptable. Why would we bother with taming a fish like cod that is slow-growing, inefficient, and ultimately not suited to live in our mangers? It is a waste of time and money.

If we must have an industrial fish, let us use a fish that works well in industrial processes and has a minimum of impact on the wild world. Both tra and tilapia are grown in fresh water, have no interaction with ocean fish, and eat primarily vegetarian feed. It is true, as the pollock industry's Jim Gilmore protests, that more and more rain forest is being cut down to grow soy for tilapia and tra feed. But more than any other form of animal protein, tilapia and tra grow extremely fast and are extremely efficient in turning feed into flesh. Much more efficient than the chickens and pigs that would end up eating that same soy if the tra and tilapia didn't.

There is one last thing that nature is telling us, and it is telling us in the form of the white, loose flake of a wild codfish—the flesh of a cod that has, as Mark Kurlansky put it, "lived a cod's life." We should be fishers of cod, not farmers of cod. And if we are to be fishers of cod, we must meet cod on their terms. To understand what cod's population dynamics are, we need to work with cod to build a long-term, stable relationship.

Humans seem to have an innate drive to master other creatures. If we must master something, in the case of the cod, instead of mastering the simplistic closed system of industrial aquaculture, perhaps we should seek the ultimate proof of our intelligence—the complete mastery and understanding of a wild system, a form of mastery in which we gradually come to understand how much fishing ground we must leave fallow as marine protected areas, areas that serve as a kind of bank-account principal from which we will earn interest every year in the form of a harvestable catch.

Let us learn how to revive the life cycles of river herring and other things cod eat. Let us learn, down to the last fish, how cod reproduce and survive in the wild and how their populations change over time. Such a mastery would include a hyperlocal fleet of knowledgeable small-scale fishermen harvesting from discrete populations of cod in as precise a way as possible. Such a fleet of fishermen might still get a small subsidy to make up for the cost of their effort, but it would be understood that any subsidy they receive is a *fee for service*, that they are stewards as well as catchers of fish, and if they fail in their stewardship role, they will lose the right to fish. In such a way, perhaps we can reconstruct a fishery where fish and fishermen are dependent upon one another, the way a flock and a herder require each other for survival.

A fish caught by a fisherman with that kind of knowledge deserves to transcend its commoner heritage. Such a fish deserves to be knighted. Such a fish should be eaten with its flesh intact, not processed by a machine and turned into a fish stick. It should not be cheap. It should be treated kindly in the kitchen, its subtle flavors and pearly flake centerpieced, and admired even if it is a little bit dull on the palate. That kind of cod I would be happy to call COD. All caps.

Tuna

One Last Bite

One of the appeals of growing up with fishing in your life is that as you mature, the possibilities expand. You can take on longer and longer trips at sea. Odd hours of the night become gradually more reasonable, and the fish you're capable of catching are tougher, harder to find, and much bigger. By the time you reach adulthood, if you still have the fishing urge and a little bit of extra cash, the whole of the ocean can become your fishing grounds. It is only in midlife that you start to reach the last frontier of your fishing adventures: your own desire to keep fishing.

In early September of 2001, I placed a phone call to the party fishing boat *Explorer.* "Thanks for calling the *Explorer,* Brooklyn's rocket ship to the tuna," the voice on the answering machine croaked. "The latest catches are: Monday night—fifteen tuna, Tuesday night—

twenty-four tuna, Wednesday night—forty-seven tuna. Tell us the date you want to go, and *Explorer* will take you to the tuna."

It is a rare party boat that hunts tuna. Nowadays it usually takes a charter boat and upward of three thousand dollars in gasoline to speed over a hundred miles into the open ocean in pursuit of them. But there is a brief window from August through November when tuna venture close enough to the coast for a handful of party boats out of New York City to seek them and make the fish catchable by the common man.

All summer I'd been thinking about tuna fishing, and I was waiting to see how the fall would shape up before committing to the trip. But one September morning it was so clear, sunny, and windless that I couldn't imagine things turning for the worse. I left a message on the *Explorer*'s answering machine booking my trip for September 28.

I was happily thinking about tuna when I left my apartment for jury duty downtown. As I turned onto Sixth Avenue in Manhattan, I noticed that the moon in the southern sky was a little more than half full. In two weeks' time it would be full. A few days off the full moon is usually the best tuna time, and I realized, happily, that I had scheduled my trip for the best possible moment. I thought about this as I let my eyes drift past the moon over to the flaming hole that had appeared in the middle of the World Trade towers.

I continued walking downtown. From a couple of miles away the hole looked big but not so big as to suggest anything more than an industrial fire of some sort. I had jury duty and did not want to get hit with the $250 fine for nonattendance. In twenty minutes I reached a gas station near the corner of Canal Street and Sixth Avenue. A large explosion rattled overhead. I turned to one of the taxi drivers filling up his tank at the gas station. "Yeah, that's 'cause them two towers is connected underground, and, like, probably it's like a gas line that blew up—that's why there was that explosion in the other tower." Sounded good to me. Didn't want to get that jury-duty fine. I kept walking downtown.

Passersby were now generally moving uptown in the opposite direction, and flecks of paper debris started to blow up the cobbled streets of SoHo. As I cut across on Lispenard Street and then merged south onto Broadway, trickles of people became clusters, which in turn started to resemble mobs. The hole in the World Trade Center that had seemed like an easily suturable scratch now appeared as it really was—a gigantic gash going deep into the heart of the building. Onward and downward I walked. Upward and northward walked the crowd. A cop finally stood in my path. "Buddy, where the hell you going?" he asked me. "I got jury duty," I said, showing him my summons and pointing to the part about the $250 fine for nonat-tendance. The cop held up his hand in the manner of a bailiff.

"Jury dismissed," he said.

I turned and joined the crowd moving uptown. By the time I reached the corner near my apartment on Eleventh Street, the World Trade towers had again become just a small silhouette in the middle distance. And then, suddenly, one dropped away and disappeared completely. Then the second vanished. And I went into my apart-ment and did not leave for two weeks.

I barely slept during that time. The only companionship I had were the tuna Web sites I had discovered on the Internet. I joined the tuna discussion boards and tracked a satellite image of a warm-water eddy drifting north, up from the Carolinas. I found myself making lifting motions, trying to push the warm patch faster so that its arrival in New York waters would match my *Explorer* fishing date.

Tuna in the western Atlantic follow the river of higher-temperature water in the Gulf Stream as it flows north from Florida up to Atlantic Canada. Eddies of warm water break away from the main stem of the Gulf Stream and wend their way to the coast within range of the sportfishing fleet. A nearshore eddy bodes well for a good trip. Conversely, no eddy means no tuna.

The eddy I was tracking did speed up, but so did the wind. Forecasts of light breezes and mild seas were revised to gales and swells. The day before my trip, I had a phone call.

"This is the fishing boat *Explorer*," an elderly woman's voice announced.

"You're canceling the trip, aren't you?" I said.

"To be honest," the voice said, "tomorrow is not looking good." A long pause. "But if you wanna go tonight, the captain—my son, that is—he says the ocean's gonna lay down."

"Is it definitely gonna lay down?"

"My son says it's gonna lay down," she repeated.

"Okay."

I filled my tuna bag with my fishing clothes, still rank from my last outing. I took every piece of food I had in my refrigerator and fried it. I put it all in my tuna cooler, loaded up my dead mother's 1989 Cadillac Brougham, and headed off from my home in Manhattan to the last serious fishing port in New York City, Sheepshead Bay, Brooklyn.

At sunset *Explorer*'s engines rumbled to life. All the fishermen gathered at the stern and stared silently into its impressive wake. The wind was picking up. Some of the fishermen were nervous, but generally there was a deep, pagan faith in the captain's predictions.

"He said it's gonna lay right down," said one.

"He said tomorrow's gonna be like a lake," said another as we began to steam out to sea.

We passed the jetty at Breezy Point and started the difficult part of the hundred-mile journey to the ocean trench called the Hudson Canyon, or, by those familiar with it, just "the Canyon." An underwater continuation of the Hudson River that dates back to the last Ice Age, the Canyon is nearly as deep as the Grand Canyon and one of largest ocean canyons in the world. Its roiling currents confuse baitfish and make it an attractive formation for tuna, bringing them in from their more distant wanderings.

The bow began to pitch. Pairs of fishermen drifted back into the cabin and shared salami, sliced by their wives "as thin as a quarter." I opened a Tupperware container and ate a lone, cold pork chop. A guy across from me read a fishing magazine, flipping through it as if it were porn. He pulled back his head, rotated the magazine to the vertical, and showed me a centerfold of a gorgeous tuna. "Hey, buddy, look at that!" he said.

The Verrazano-Narrows Bridge winked out of sight, and *Explorer* took on a heavier roll. An extremely fat man laid himself out supine on the deck and began a wet, blubbery snore.

"They said it was gonna lay down," the guy with the fishing magazine said. "They said, like, five- to-seven foot waves. By the time we get out to the Canyon, that sea's gonna be ten to twelve."

A massive cooler slid across the floor and hit the extremely fat man in the gut. He swatted it away and continued snoring.

"I wasn't gonna go," the fishing-magazine guy continued. "I knew it was gonna be like this. But I walked out of the Trade Center a couple weeks ago. Tower Two. My name's Steve, by the way," he said, offering his hand.

"Paul," I said, offering mine.

"Haven't we fished together before?"

"Maybe," I said. "Do you ever go on the *Helen H*?"

"Sometimes," he said. And then we were both embarrassed and silent. I went belowdecks.

I can't say that I slept, exactly. I stole little patches of unconsciousness during the boat's slow uplifts as we plowed farther and farther away from the realm of men into the realm of tuna. In the weightlessness of one crest, I felt my head grow light with the feeling of impending disaster. Then we went into free fall down the other side of the wave and boomed into the trough. My head bounced up and hit the hard metal ceiling.

Three hours of this drove me topside. I climbed up the stairs and sat on one of the benches and tried to deal with my tuna gear. I started tying tuna hooks onto strands of fluorocarbon tuna leader. Staring at the knots while the boat ground against the waves was nauseating, but I persisted because it was necessary. On one of the fishing sites I had started frequenting, an article called "The Holy Tuna Tablets" maintains that fluorocarbon tuna leaders "give you the edge." Tuna can direct heat to their huge, luminous eyes, giving them much better vision than most fish and allowing them to see most commercial monofilament fishing line in the water. But modern fishermen have dissected the adaptations of tuna and in each case come up with a conquering strategy. Fluorocarbon line is a polymer that's almost invisible to fish because it refracts and bends light at nearly the exact angle as water.

Once the fluorocarbon leaders were tied, I took another Dramamine and waited for the tightness in my throat to back off from a full-scale vomit. Then I started fixing up my squid rig. "The Holy Tuna Tablets" advises you to bring a squid rig to catch your own live squid for bait. During the previous year's tuna trip, when the tuna were eating only squid, I didn't have a squid rig. I asked one of the guys at the rail if I could use his.

"Get your own fucking squid rig," he said.

At around two in the morning, *Explorer*'s engines ratcheted down, and Steve, the guy who had walked out of Tower Two, walked out of the cabin and joined me at the rail. "How ya doin' there, Paulie?" he said.

"Not so good, Steve."

The wind seemed to be coming from three sides now, but there was nothing to do except fish. Out here in tuna territory, the televisions in the cabin were dead and nobody's cell phone had reception. I thought that once we anchored, I wouldn't have such an overwhelming urge to vomit, but anchoring made it worse. Without any forward movement, the wave period had no predictable sequence, and in the deep-space blackness of night on the Canyon there wasn't even a horizon to help you get your bearings. I tried my squid rig. There were no squid. I pried a piece of bait out of a half-frozen mess of butterfish and threaded the hook into the fish's gullet. One of the mates ladled out butterfish chum. Flakes of butterfish flew in the wind and stuck to my face.

"This sucks," said Steve of Tower Two.

The wind gusted harder. The sea looked cold and barren. Every few minutes the captain called out sonar readings over the staticky PA system:

"We got a nice bunch of fish at a hundred feet, guys." *Pshht.*

"They're way down, now, down at two hundred feet, guys." *Pshht.*

"They're up at eighty feet now, guys." *Pshht.*

The ghosts of tuna passed unseen beneath our feet. The

butterfish-chum slick was punctuated by human vomit now as the less experienced fishermen started to experience their evening meals for a second time. The extremely fat man came off the cabin floor, walked to the rail, and expelled a tremendous vomit, in proportion to his great size. He went back inside and laid himself back out on the floor.

"Fish are back down at a hundred feet, guys, *pshht*," the captain said over the PA.

I peeled twenty feet of line off my reel. I closed my eyes. I imagined my butterfish fluttering down—a dull little silver coin falling into the empty black purse below. It was a nauseating image. I opened my eyes again. My reel's spool was spinning on its own. Spinning much faster, in fact, than the weight of a falling butterfish should have allowed.

"Hey . . . hey . . . buddy . . . Paulie!" Steve shouted. "Hit 'em!"

He reached over and flipped the spool lock on my reel. I was slammed down onto the rail.

"Yeahh!" shouted Steve. The first tuna of the night was on my line.

Like the *Explorer*, tuna are all engine. On the line they feel like no other fish, and one can almost imagine the Schwarzenegger-type muscles, flexing and pulsing their myoglobin-rich tissue in coordinated, punishing synchronization. They are the one fish out there that make a fisherman think, "I don't know if I can do this." As the tuna sprinted off I sank into a kind of squat, trying to shift the stress from my back onto my knees.

This in turn caused a problem. I had worn extra-wide pants to accommodate long underwear. But the weather was downright balmy out in the Canyon, and I had shed my long underwear before coming to the rails. Now, as the tuna surged forward and I bent my knees in a defensive crouch, my baggy pants fell down to my knees. When I reached around to pull them back up, the tuna seemed to sense it and swam even harder. My pants finally settled midway down my thighs.

Steve came up behind me. "You don't look so good," he said.

"I . . . know."

"You want me to take that pole for you?"

"No . . . I'm . . . fine," I said.

"You don't have to feel ashamed," he said. "Tuna are tough."

"I know."

Ten minutes into the fight, the tuna stopped swimming and I stopped reeling.

Steve stood by my side and shook his head. "Jesus," he said.

Then all at once it turned and charged the boat.

"Reel, buddy, reel! Reel, reel, reel!" Steve shouted.

I put my head down and cranked, trying to take up the slack.

"Jesus, buddy, watch out!" I heard Steve cry. I opened my eyes and saw that the hood strings of my sweatshirt were dangling in my reel. I was about to be bound Ahab-like to my fishing pole. Gingerly, Steve reached around my shoulders and tucked the sweatshirt strings into my collar.

"Thanks, man," I said.

"Doin' good, buddy."

The tuna began swimming a broad, slow arc around the boat.

"It's the death spiral," Steve said solemnly.

The death spiral's diameter decreased with each big circle. Three more circles and I dared a peek over the rail. Way down, just beyond the glow of the boat's running lights, I saw a muted green flash.

"Hey," Steve called out, "we got color here."

A mate came by and looked down into the water. "All that screaming over that little thing?" he said.

The mate lowered the gaff. I held my breath. Sometimes when tuna see the shining metal hook coming at them, they streak off in a ridiculous, last-ditch escape run, sometimes tail-walking across the water before busting the line. But the mate dipped down agilely and struck into the fish's flesh. He tried to lift it into the boat.

"Jesus," he said.

"Not so little?" I asked.

"Not so little."

The mate was yanked onto his tiptoes by the unseen force below. Steve thought fast and grabbed another gaff and struck. He and the mate swayed back and forth like a pair of girls doing the hula. Then they each exhaled, counted to three, and raised their gaffs hand over fist in sync. The fish rolled over the rail, furious. Steve shook his gaff free and jumped back. The tuna hit the deck hard. It was a yellowfin tuna, and, true to its name, its dozen-odd finlets running down the aft ridges of its torso glowed canary bright in the darkness. In spite of its hard fight, the fish was still angry and dangerous and as big as an adolescent. Its huge, superheated eye met mine, and we both gasped for air.

If the tuna had had a voice and the power of reason, it would have screamed and pleaded at this point. But the only expressive thing about a tuna is its tail. All a tuna ever does its whole life is crank its tail back and forth with great determination. Even after it's caught, after it's in the air, it never occurs to a tuna to switch off its relentless tuna motor. *Bap-bap-bap-bap-bap*, it goes until the mate cuts its throat and the blood pours out onto the deck. *Bap-baap-baaap-baaaap*—the engine runs down and then stops cold.

"Congrats," said Steve.

"Thanks," I said, and vomited.

The forty-eight species of the family Scombridae that are known collectively by the word "tuna" are among the fastest, most powerful fish in the world. One derivation of the word "tuna" goes back to the Greek verb *thuno,* meaning "to dart," an ascription that suggests the perspective of a primitive angler caught flat-footed while the fish shoots away into oblivion. In fact, tuna do much more than dart. Contemporary biologists have cruised alongside them as they breach, dolphinlike, in the water, clocking them during their fastest accelerations at speeds in excess of forty miles per hour, faster than the *Iowa*-class battleships, the fastest warships ever built. Whereas battleships require a constant fuel supply, the largest tuna—the Atlantic bluefin—covers distances that stretch across the foodless depths of the midocean trenches, journeying from the Mediterranean Sea to the Gulf of Mexico in the Atlantic. Their range encompasses nearly the entirety of the ocean.

Even the most confirmed enemy of "intelligent design" theories can have a hard time imagining the forebears of these great fish inching slowly down an epochs-long evolutionary course to become modern tuna. They seem like deus ex machina incarnate or, rather, machina ex deo—a machine from God. How else could a fish come into being with a weird slot, as hard and fixed as the landing-gear slot on an airplane, into which it retracts its dorsal fin to achieve faster speeds? How else could a fish develop a whole new way of swimming where a slim crescent of a tail, insignificant in size compared to most fish tails, vibrates at astronomical speed while the rest of the body slips forward with barely any bend, pitch, or roll? And how else would a fish appear within a phylum of otherwise cold-blooded animals that can redirect the heat that its muscles throw off back into its very flesh and raise its body temperature by as much as twenty degrees above ambient conditions? Yes, the biggest tuna are warm-blooded.

That tuna can be extremely large—in excess of fourteen feet and fifteen hundred pounds—is just one side note of how exceptional they are. For those of us who have seen their oversize-football silhouettes arrive, stop on a dime, and then disappear in less than a blink of an eye; for those of us who have held them alive, their smooth hard-shell skins barely containing the surging muscle power within, they are something bigger than the space they occupy. All fish are a distinctly different color when alive than when dead on ice in a seafood market. But with tuna the shift from alive to dead is orders of magnitude more profound. Sometimes fresh out of the water with their backs pulsing neon blue and their bellies gleaming pink-silver iridescence they seem like the very ocean itself.

And in a way they are. If salmon led us out of our Neolithic caves in the highlands down to the mouths of rivers, if sea bass and other coastal perciforms led us from the safety of shore to the reefs and rocks that surround the coasts, and if cod and the gadiforms led us beyond the sight of land to the edges of the continental shelves, tuna have taken us over the precipice of the continental shelves into the abyss of the open sea—the final frontier of fishing and the place where the wildest things in the world are making the last argument for the importance of an untamed ocean.

Fish like cod and striped bass with populations contained within discrete national boundaries have in a few cases begun inching their way back toward viability. Effective management measures can be put in place because the people who fish those stocks generally recognize the regulating authority as legitimate and understand that there will be real financial impact on their lives should they transgress.

Tuna, however, range over open ocean and cross multiple nations' territorial waters. They are thus now subject to what regulators call an international agreement but what environmentalists might label a free-for-all. In the last fifty years, as humans have outstripped their coastal fisheries and advanced their fishing technologies, fishing has moved out of national territorial waters into what is known in nautical parlance as the "high seas"; these areas are owned by no one and fishable by anyone.

Catches from the high seas have doubled in the last half century, and much of that catch increase has come in the form of tuna. Moreover, because tuna cross so many boundaries, the way international tuna treaties are set up means that even when tuna do tarry in any one nation's territory, they are still technically catchable by any other treaty member nation. The conventions that govern tuna allow any tuna-fishing nation to fish in any other tuna-fishing nation's waters, provided the fishers stay within an overall quota of fish caught—a quota that no nation seems to have the resources or the attention span to adequately monitor and enforce.

On top of all these regulatory challenges, there is the rise of sushi in the past three decades and the new demands that this phenomenon has put on tuna stocks. Curiously, tuna sushi is a relatively new invention, even in Japan. As Trevor Corson, an East Asia scholar and author of *The Story of Sushi*, wrote me recently, the cultured palates of the Japanese aristocracy generally preferred delicate white-fleshed snappers and breams over heavy, red-fleshed tunas. "Many of the so-called 'red' fish were thought to be too pungent and smelly," Corson wrote, "so in the days before refrigeration, discerning Japanese diners avoided them." All this began to change in the nineteenth century, when an abundant catch of tuna one season prompted a Tokyo street-stand sushi chef to marinate a few pieces of tuna in soy sauce and serve it as "nigiri sushi." The practice caught on. Generally, smaller, leaner yellowfin tuna were used for nigiri. Occasionally a big, fatty bluefin came to market, but as Corson pointed out, these big bluefin were nicknamed *shibi,* or "four days," because chefs felt they had to bury them in the ground for four days to ferment and mellow the heavy, bloody taste of the meat. From the few stalls that served it in the Edo period, tuna caught on and by the 1930s was considered an integral part of a good sushi meal.

At first, Japanese tuna fishing was relatively contained. As a term of Japan's surrender at the conclusion of World War II, Japanese vessels were prohibited from fishing beyond their territorial waters throughout the 1940s. But when the prohibition was lifted in 1952, things started to change. As Dr. Ziro Suzuki, an authority on Japanese offshore fisheries wrote to me, "In order to recover from the devastation of the war, Japanese fishermen needed more tunas to secure food for domestic demand and also to earn more money by exporting tunas to the canning industries in Europe and the U.S." When the technology to deep-freeze tuna in the holds of fishing vessels was invented in the 1970s, though, more and more tuna could be served raw rather than canned. Suddenly fishermen could haul in tuna from the farthest reaches of the oceans, freeze them immediately, and keep their catch sushi-ready for as long as a year. Tuna sushi was suddenly exportable.

The evolution of the Western/Japanese sushi relationship had other twists. In the late 1960s and early '70s, Americans and Canadians ramped up the sportfishing of giant, thousand-pound Atlantic bluefin tuna, principally off Canada's Prince Edward Island and Nova Scotia. Most of these fish were caught, killed, and then discarded at the town dump, for, just like the Japanese, Americans considered bluefin too bloody to eat and had no interest in bringing home their catch. But the bluefin sport fishery developed at the same time as the Japanese export boom to North America. Cargo planes from Japan, stuffed with electronics and other consumer goods, would arrive in American airports only to fly back empty to Japan—a huge waste of fuel. It was only when several Japanese executives realized that they could buy bluefin for pennies on the pound from American sportfishermen that they began filling empty cargo holds with bluefin and flying them back to Japan. Within a few years, Japanese began esteeming bluefin above all other tuna, and this fetishization boomeranged back to the West, which soon developed its own bluefin appetite.

The West's embrace of the Japanese sushi tradition had another multiplier effect: it brought people who had previously disliked fish into the fish-eating fold. I saw this immediate effect within my own family when my brother moved to Los Angeles to become a screenwriter. "You know how I've always been about cooked fish," he wrote me when I asked him about his newfound sushi habit. "I couldn't stand the smell or the taste or the texture. The few times I had to eat fish were usually at dinner parties. In those cases, I would breathe through my mouth so I couldn't smell it and swallow small pieces whole so I wouldn't have to taste it.

"Okay, cut to 1992," my brother, the writer of horror films like *Halloween H20* and the Stephen King adaptation *1408*, continued. "I'd just moved to Los Angeles. After a lot of peer pressure, I finally agreed to go to our local sushi restaurant to try some. I ordered a regular tuna roll, thinking I would do my hold-my-breath-and-swallow-it-whole thing. But when it arrived, I immediately noticed something different—it didn't smell 'fishy.' I dipped a piece in soy sauce mixed with a little wasabi, put the damn thing in my mouth, and chewed. Man, it was like that great moment in the film *Ratatouille,* where the evil food critic tries the eponymous dish and is suddenly transfigured. The raw tuna tasted nothing like cooked fish. Pun intended, I was hooked."

What my brother tasted was a biochemical phenomenon that can be experienced with many high-speed, fatty fish but which is particularly true of tuna. Hard-swimming fish like tuna use large amounts of a chemical called adenosine triphosphate (ATP) to store and expend energy. After death ATP is converted to inosine monophosphate (IMP), a chemical associated with the "fifth" flavor Japanese call umami, or "tastiness." It is a flavor that even non–fish eaters find pleasant on the tongue. When cooked, however, IMP breaks down and combines with other chemicals present in fish flesh and produces flavors that people like my brother find unpalatable. In addition, the odors that might be emitted by not-so-fresh fish are neutralized in Japanese sushi techniques by soy, ginger, and wasabi.

The global rise of sushi combined with the international failure to formulate a functional multination fishing agreement around tuna has led to progressive declines in many tuna stocks, the worst of which has been the decline of the two intermingling stocks of Atlantic bluefin tuna: the Western stock, which spawns in the Gulf of Mexico, and the Eastern stock, which breeds in the Mediterranean. Atlantic bluefin are the biggest and slowest-growing of the tunas; the Western stock can take more than seven years to reach sexual maturity and considerably longer to become "giants"—that is, the five-hundred-plus-pound spawners that many biologists feel are the key reproductive engines of bluefin populations. Since it is the giant spawners that are the primary targets of exploitation, their numbers have crashed. Fishermen always appreciated giant bluefin as animals—as fighters on the line or as evaders of harpoons. But the rapid growth in giant bluefin price, from pennies to hundreds of dollars a pound, created a different kind of appreciation. Today the passion to save bluefin is as strong as the one to kill them, and these dual passions are often contained within the body of a single fisherman.

"I love these fish," a commercial bluefin-tuna harpooner told the reporter John Seabrook in a 1994 issue of *Harper's Magazine*. "But I love to catch them. God, I love to catch them. And I know you need some kind of catch limits because I'd catch all of them if I could." As bluefin get more and more valuable on the marketplace (prices for a single fish have topped $150,000) the commercial fishermen who pursue them get more and more twisted in their behavior—a bit like Tolkien's Gollum pursuing the ring. They are endlessly attractive to fishermen when present but can leave fishermen holding massive bills for fuel, bait, and gear when they vanish. It sometimes seems as if a Gandalf of fisheries management is needed to work up an incantation that would save the fisherman from the destructive relationship he has with the great fish, the one that tempts him to destroy the very profession that would sustain him.

The bluefin conservation advocates, often former tuna fishermen who have been able to pull themselves away from the lure of tuna's silver-ingot bodies and marbled-sirloin flesh, have tried all manner of spells to get those who eat tuna or those officials who legislate over them to somehow sit up and take similar notice—to abstain from eating them or to pass enforceable regulation for the sake of their preciousness. It is this often-futile battle that is the most telling part of the tuna fishery today. It is the battle with ourselves. A battle between the altruism toward other species that we know we can muster and the primitive greed that lies beneath our relationship with the creatures of the sea.

And yet it is a battle that has been fought and won before, against high odds. Looking back over the history of the ocean, we can see that there is one order of sea creatures bigger than tuna that has earned our empathy and, more important, our protective resolve, rising up from the background of marine life to become a superstar of conservation, on a par with the tiger and the elephant. It is to this example we must look if we are to fix our tuna problem once and for all.

W hale Carpaccio—130 Kroners." Thus read the lead appetizer on the menu before me in an upscale Norwegian restaurant where I was dining on a recent winter evening. Eight slices of whale arranged raw on a plate for the reasonable price of about twenty U.S. dollars. I have to admit that the prospect of ordering it was intriguing. I had never been to a country that still practiced whaling, and I had certainly never seen whale on a menu. What would whale taste like? I wondered. Would it be fatty and chewy like beef, or would it have the loose, flaky texture of fish that don't need dense muscles to resist the pull of gravity? Would it be served like prosciutto, with a thin slice of Parmigiano-Reggiano cheese? Or, since carpaccio is an Italian dish and Italians avoid mixing cheese and seafood, would it be more appropriate merely to drizzle olive oil over the whale's buttery sheen?

These were the thoughts that made my mouth water as the waitress approached my table. But when she took the pen from behind her ear and asked me in blunt Nordic style if I'd like to "try the whale," all at once my twenty-first-century foodie curiosity wilted. "No," I said, "I'll have the mussels."

I would like to be able to say that I did not "try the whale" because of some superior moral quality I possessed. But which animals we think of as food and which we think of as living creatures is highly contextual. My conception that a whale was somehow too good to eat comes from a historical process that predates me by nearly two centuries, a process that has yet to happen with fish.

Up until 1756, when the French zoologist Jacques Brisson published *Le Regne Animal Divisé en IX Classes* (*The Nine Classes of the Animal Kingdom*), whales were thought of by both the scientific and the lay communities as just very big fish. It was only when the father of taxonomy, Carl Linnaeus, in the tenth edition of his *Systema Naturae,* confirmed Brisson's definition of whales as nonfish that a certain ennoblement of those animals started to occur. When Linnaeus went a step further and classified whales as mammals, that rankled Brisson, who suspected Linnaeus of overstepping the realms of scientific acceptability and trying to disguise his plagiarism of Brisson's findings in an outlandish hypothesis. But Linnaeus was steadfast in his beliefs, asserting that whales "by good right and just title according to the law of nature" deserved to be classed with mammals.

Mammalian or not, the fact that "cetaceans," as the group would later be named, were not fish was well established within the scientific community by the end of the eighteenth century. By the new century's turn, the idea that they were fish would start to sound downright preposterous to any self-respecting scientist. "This order of animals," the English zoologist John Hunter wrote in the early 1800s, "has nothing peculiar to fish except living in the same element." Yet even with Linnaeus's and other major scientists' imprimatur, it took many decades for the general public to accept the fact that whales were different and somehow special. Nowhere was this more evident than in the 1818 New York trial of *Maurice v. Judd.*

Brilliantly reexamined by D. Graham Burnett in his 2007 book *Trying Leviathan,* the *Maurice* affair exposed the broad rift that existed and still exists between the measured findings of scientists and the "common sense" of the everyday consumer. By all rights the facts of the case should have buried it forever within the mountains of records in the New York legal system. The suit came about because of a new development in seafood regulation. Early in the nineteenth century, the state legislature of New York began requiring that all fish oil be inspected in order to allow better grading and to reduce the tendency of oil merchants to disguise one type of fish oil as another. In the case of *Maurice v. Judd,* Maurice (an inspector) had fined Samuel Judd (a candle maker) seventy-five dollars for buying three barrels of fish oil that had not been inspected. Judd refused to pay the fine, insisting that he had bought not "fish oil" but rather "whale oil" and that furthermore whales were not fish.

This somewhat trivial dispute might have been banged away by the thud of a less patient judge's gavel, but instead the trial became a media circus, in part because of a colorful array of witnesses that included a whaler by the name of Preserved Fish, but mostly because of the participation of Samuel Mitchill, an Enlightenment naturalist and New York City's most famous scientist. For two days Mitchill sparred with William Sampson, a respected and often wily prosecutor, trying to establish the great differences between whales and fish: That whales were warm-blooded. That they breathed air. That they lacked scales. And, in a moment that shocked the standing-room-only crowd, Mitchill declared that "a whale is no more a fish than a man."

But in spite of Mitchill's depth of knowledge and his standing in New York society, his careful explanations ended up confusing and even enraging the trial's jury. After a short deliberation, the jury returned a verdict that slapped down a hundred years of careful scientific investigations. A whale, the jury foreman announced to the assembled crowd of journalists, gossipmongers, and wharf dwellers, *was* in fact a fish.

The press taunted Mitchill for days afterward. "Pray sir, how goes it with whale oil now?" wrote New York's *Evening Post*. "Is it oil of fish, or of flesh, or of *red herring*?"

Yet while Mitchill's reputation certainly suffered after the humiliation of the trial, the whale's standing began to rise. The trial lingered in the popular unconscious, and the first inklings that suggested that the whale merited exceptional consideration began appearing in print. A sperm whale's ramming and sinking of the whale ship *Essex* (the inspiration for the novel *Moby-Dick*) in 1820, just two years after *Maurice v. Judd*, gave the impression that whales had agency and had identified humans as their enemies. And in the 1820s, the common practice of harpooning a whale calf and waiting for mothers to gather around it so that multiple kills could take place was criticized publicly. The idea that whales even had intelligence was broached. In *Trying Leviathan*, Burnett relates how a book called *A Whale's Biography* came out in 1849 and the following year in Honolulu the newspaper the *Friend* ran a letter to the editor from "Polar Whale," address "Anadir Sea, North Pacific," in which the cetacean writer identified himself as hailing from "an old Greenland family" and "pleaded for 'friends and allies' to 'arise and revenge our wrongs' lest ignominious extinction descend upon his 'race.'"

But these flashes of sympathy were negligible compared to the ruthless expansion of the whaling industry and the effect that expansion had on a naturally sensitive order of animals. It is an essential truth of ecology that big animals tend to be the scarcest because of the scope of resources they must command. They are kings of sorts, considerably less numerous than commoners. Local populations of whales were therefore easily extirpated. And when they were, whaling fleets journeyed to the far extremes of the globe in search of untapped schools. When even those far-flung populations started to show declines, humans may have for the first time gotten a glimpse of their destructive potential. Whereas once the seas seemed inexhaustible, the decline of whales on a global basis showed that it was indeed possible to overexploit the oceans and drive a creature (and an industry) to commercial extinction.

Ultimately, though, it was not just whales' scarcity that led to the end of the first era of whaling; rather, it was the appearance of cheaper, more easily obtainable whale-oil substitutes that changed the rules of the market and spared the remnant populations of "oily" whales. Petroleum oils made sperm-whale products commercially irrelevant long before they were made illegal in the early 1970s.

What is more significant, though, to a discussion of the future of fish, particularly big fish like tuna, is what happened during what is known as whaling's second era—the time during which humans moved from using whale oil to light their lamps to using whale oil and other whale parts for a much wider scope of applications, including fertilizer, lipstick, brake fluid, and even human food.

This second phase of whale exploitation began in the 1870s. During this more aggressive phase, steam- and later diesel-powered vessels, explosively launched harpoons, and compressed air flotation devices were developed that enabled whalers to hunt a whole new range of very large, even more naturally scarce species. Before these inventions came along, blue whales and fin whales were too fast to catch and would sink to the bottom if killed. After these technological breakthroughs, the largest creatures ever to live on earth were killed instantly with exploding grenade tips, secured to artificial flotation devices, and brought to market—giant rafts of fat and protein fit for a range of industrial purposes. Postwar Europe was particularly drawn to the use of whales. Indeed, if you are from Europe and born before 1960, no matter how much of an environmentalist you may consider yourself, there is a high likelihood that you have eaten whale. In the 1940s and '50s, while European agriculture was still recovering from World War II, whale fat was regularly rendered and put into margarine and other oil-requiring foodstuffs. Even if you abstained from margarine, there was a good chance that whale still made it into your body indirectly—the meat and bones of whales were used as fertilizer to grow vegetables.

There were even bigger plans for whales. Increasingly, as postcolonial unrest and Cold War competition for favor in emergent African, South American, and Asian nations grew, significant attention was paid to figuring out a way to alleviate hunger in those countries. Before the agricultural advancements of the Green Revolution came online, economists feared that the world was on the brink of a Malthusian collision of population growth and food shortages. Some agronomists suggested that whales could become a significant protein source for the impoverished Third World. Collaborations between nuclear scientists and marine biologists were even proposed whereby tropical atolls, blown up by nuclear testing, could be used as giant corrals for the commercial farming of whales.

But all this optimism about the potential of whales was quickly checked by the reality of the drastic decline in whale populations. By the 1930s, cetacean numbers were so low as to provoke three successive international agreements. By the time these agreements were raised up to the level of a convention in 1947, participating nations deemed that these measures were necessary in order "to provide for the proper conservation of whale stocks and thus make possible the orderly development of the whaling industry."

It is important to note here that nowhere in the convention that led to the creation of the International Whaling Commission was there mention of conservation of whales for the sake of their specialness. Rather, as with all things from the sea in earlier days, conservation was seen as necessary for the sake of future exploitation. Just as opposition to slavery was once argued from an economic point of view, the antiwhaling movement had its origins in financial motives. Like abolitionism, it had to develop a second, *moral* prerogative to spur an appropriate response from man. And, as with abolitionism, the debate took place around the issue of intelligence.

The beginnings of whale conservation for the whale's sake alone dribbled out of studies of whale communication. Dr. Roger S. Payne is a Harvard-trained biologist who did his preliminary work on echolocation in bats and owls. But a desire to enter the field of conservation biology led him to apply to the sea what he had formerly studied on dry land. "I wasn't doing anything that was directly related to problems that I, as a biologist, am deeply and bitterly aware of," he recalled, "which have to do with the destruction of the wild world by people. So I thought, if all you've had in training is the chance to work on the acoustic worlds of animals, what animal could I work on that needs my help?"

Along with the researcher Scott McVay, Payne began studying the "songs" of humpback whales and over time developed a theory that not only do whales communicate with one another in a complex and ever-evolving way, but that in certain species (blue whales and fin whales) whale song could be transmitted across the entire breadth of an ocean. The theory was alternately embraced and attacked within the scientific community, but in 1970, when Payne released an LP entitled *Songs of the Humpback Whale*, the public itself became the greatest adjudicator of whether or not whales were intelligent. The album, which was followed by a sequel that included jazz "duets" between a humpback whale and the saxophonist Paul Winter, sold 10 million copies and today is still the biggest-selling wildlife record-ing of all time. Payne's whale-song recordings went on to underscore numerous popular ballads, including John Denver singles and David Crosby and Graham Nash's 1975 *Wind on the Water* album.

Other researchers, like the New Zealander Paul Spong, who had developed a series of communication experiments with two killer whales named Skana and Hyak, brought the whale-conservation movement into the political arena. Spurning the classic noncommit-tal stance that scientists are supposed to take in relation to politics, Spong banded together with early anti-nuclear-testing activists and helped form Greenpeace. According to Rex Weyler, an early Green-peace member and unofficial historian of the organization, it all came down to the little inflatable boat and the big Soviet whaling ships. "I would suggest," Wexler wrote me, "that the precise moment the Save the Whales movement entered the popular unconscious would be July 1, 1975." On that day images of a small Greenpeace Zodiac poised between a breaching whale and a Soviet commercial whaling vessel in the Southern Ocean were broadcast around the world by Walter Cronkite on CBS News and on other major media outlets. That image remains to this day the enduring image of the Save the Whales movement and one that helped to create a substantial conservation lobby within the International Whaling Commission.

Even though the International Whaling Commission was intended to be a fisheries-management organization, over the course of the seventies and early eighties environmentalists pushed it into playing an international environmental-protection role. This was ultimately codified in a landmark agreement that took place in 1982. For the sake of those countries that were only grudgingly brought to the whaling negotiating table, the agreement was given three years to be put into effect and expressed not in environmental or ecological terms but rather in the language of fisheries management.

From 1985 forward, the International Whaling Commission declared that throughout the world there was to be a "zero catch quota" for whales. In other words, a complete and total world moratorium on hunting an entire scientific order of animals. It was the broadest and most far-reaching act of kindness humanity has ever bestowed on another group of species. Though contested and embattled and fraught with disagreements that result in violations, this kindness has persisted. The whaling moratorium remains in effect to this day.

In the summer of 2006, an editor from the *New York Times'* opinion page e-mailed and asked me if I would like to write a short article on whether people should continue to eat fish. "We were wondering if you'd be interested in writing an essay whose basic argument would be that we should get our Omega3's from somewhere else and just not eat fish anymore because of all the problems fishing causes," the editor wrote. "Or, alternately," she continued, "an essay arguing that we shouldn't feel guilty eating fish despite all the problems it causes."

I considered this question for a long time and conferred with many people on both sides of the issue. In the end I decided to try to track a middle course, saying that yes, we should still eat fish, that it was important that we still regard the ocean as a living source of food and not just a place to dump our garbage. However, I stipulated that a few basic guidelines should be followed to find a balance between human desire and ocean sustainability. I covered the usual topics one comes across at sustainable-seafood conventions: That one should favor fish caught by small-scale hook-and-line fishers because of the lower impact on sea beds and underwater reefs. That when choosing aquacultured fish one should choose vegetarian fish, like tilapia and carp, because of the lower strain they put on marine food webs. When it came to tuna, though, I offered no triangulation whatsoever, because in my view there simply was no compromise possible. "Don't eat the big fish," I declared toward the end of the editorial. "Dining on a 500-pound bluefin tuna is the seafood equivalent of driving a Hummer."

But two weeks after making my high-minded pronouncements, I found myself at a family dinner party at an upscale Manhattan restaurant. The appetizer choice on the prix fixe menu was either a mini–sirloin steak or bluefin tuna carpaccio. It would seem the choice should have been simple. I had my principles, and I had expressed them quite publicly. But unlike the earlier moment in Norway when I successfully kept myself from ordering whale carpaccio, this time, nearly without hesitation, I chose the bluefin. I quickly scarfed it down and nearly forgot about the delicious paper-thin slices after they had been washed away with a glass of pinot grigio. I turned to my twelve-year-old daughter, who had ordered the sirloin steak, and asked her how her food was. She had just read my *New York Times* op ed in draft form. "Hypocrite," she said coolly.

In my feeble defense, I am not alone among seafood writers in my sampling of bluefin. In several of the fish-in-danger books I've read over the course of researching my own fish-in-danger book, the intrepid author inevitably visits Japan's vast Tsukiji fish market and marvels at the giant carcasses of bluefin tuna being put up on blocks for auction. After condemning the trade, the author in question slips into a side stall and has one last delicious bite of bluefin tuna before promising never to eat it again. My guess is that if these authors were served whale carpaccio, they would, like me, have no trouble refusing. In the modern world, whales are simply not considered food, while bluefin tuna is judged an acceptable delicacy.

But based solely on numbers, the whale carpaccio is the carpaccio of choice. The whale on my menu at that restaurant in Bergen, Norway, was most probably a minke whale, an animal whose population in the wild, after the moratorium was put in place, has grown to over a quarter million animals (estimates vary widely, with some putting the population at close to a million). Norwegians have partially withdrawn from the whaling moratorium and now conduct "scientific whaling" for research purposes. Some of those research subjects end up in restaurants as whale carpaccio.

But judging Norway's dubious moral position vis-à-vis whales becomes increasingly problematic when those countries doing the judging are nations that fish bluefin tuna. The most pessimistic estimates indicate that the population of North Atlantic bluefin may have already imploded beyond the point of recovery. Some scientists place the total number of giant bluefin spawners in the Western stock of the North Atlantic at a mere nine thousand animals, or, in food terms, about 43 million individual slices of sashimi—enough for every adult American living along the bluefin's migration route to have one last bite.

Every year, though, fishing for bluefin tuna continues to grow. In 2007 every single member nation violated catch limits set by a kind of International Whaling Commission of tuna, called the International Commission for the Conservation of Atlantic Tunas (ICCAT).

Bluefin, then, represent a very whalelike dilemma. They are big. As such, they are ecologically limited in their populations. Even before they were commercially fished, they were never anywhere near as plentiful as cod, salmon, or sea bass. Like whales, whose migrations carry them sometimes from pole to pole, tuna are far-ranging and committed to no single nation. Their transience is intractable and indeed imperative to the continuation of their life cycles.

They are in all respects an unmanageable fish.

International regulators have as much as admitted their inability to manage the species. In a recent report on the status of the North Atlantic bluefin tuna stock, ICCAT officials wrote from Madrid, "Based on the Committee's analysis, it is apparent that the catch limits set by ICCAT are not respected and are largely ineffective in controlling overall catch." It concluded, rather bizarrely, that "the current management scheme will most probably lead to further reduction in spawning stock biomass with high risk of fisheries and stock collapse."

In other words, ICCAT apparently believes that its own management of bluefin tuna is contributing to the fish's demise.

Other, smaller tuna—the longfin albacore that makes up the bulk of the canned-tuna fishery, the yellowfin and bigeye tuna that are often labled as "ahi"—may be more manageable—they grow faster, spawn earlier, and have less far-flung peregrinations. But if the bluefin goes bust, the other species are next in line. They are sure to face increasing fishing pressure from a broader range of nations.

The reasons for the ever-expanding pressure on tuna go back to the original problem with fish, fishermen, and fisheries. Fishing is still governed by primitivism rather than by rational thought. Humans today might be organized into nations with treaties, international negotiation tools, lawyers, and formal protocols, but the essential dynamics of tuna fisheries are akin to the dynamics among members of a hungry tribe surrounding a carcass. In prehistoric times, in periods of food scarcity, individual members of a tribe of humans might have fought over the different chunks of flesh cut from a single animal. In modern times the fight is the same, only the individual hunters are nations and the carcass is an entire population or species of fish. And the term that is used by nations arguing for their rights in this primeval tuna struggle is "fairness."

Joseph Powers is a former chair of the scientific committee for ICCAT and currently a professor of fisheries science at Louisiana State University. As someone who has followed the tuna debate for decades, he has seen a consistent argument occur over and over again during negotiations. "In the tuna debate, there are a lot of historical dynamics that date back to colonialism," Powers told me. "When you start talking about negotiating quotas, the first thing that comes up is the historical catches of rich countries. People from developing countries in Africa where a lot of tuna fishing happens will say, 'You came down and nailed us over the years, and so we're entitled to catch as much now as you had back when tuna fishing started.' And so you get countries like Brazil, Namibia, and the North African countries all wanting their piece of the action." Typically at ICCAT negotiations, representatives from developing nations come to the table highly aggrieved. They have in their hands historical catch statistics that very clearly show that First World nations like Spain, France, and Japan have caught a lot of tuna. But now that African and South American nations are advanced enough to put their own fishing fleets in the water, they feel that they should get as much fish as the developed nations once caught in earlier times. This despite the fact that there simply aren't enough fish to satisfy the abstract terms of "fairness." As Powers sees it, "Even though scientific advice says you should stick to a specific catch number, in order to negotiate a deal they tend to nudge that number over a little bit." That little move to the right or the left is enough to put a population of tuna in jeopardy.

For people who have seen the scientific evidence ignored year after year, particularly with bluefin tuna, it is the feeling of watching Ted Ames's imbecile making that same mistake again and again and never learning. And now advocates trying to save the bluefin have realized that in this atmosphere of tribalism, getting nations to reach a consensus that is in line with scientific evidence is impossible.

Bluefin-tuna conservationists have therefore opened up a second front. In the last two decades, they have pushed multiple times to list the bluefin with the Convention on International Trade in Endangered Species, or CITES, a status achieved by tigers, rhinos, and whales and a change that would end the international trade in bluefin, theoretically putting an end to its export to its primary market, Japan. But every time this option is put forth the internal squabbles of the tuna-hunting nations prevail. At the most recent CITES convention, in the spring of 2010 in Qatar, for the first time both the United States and the European Union backed a CITES listing for bluefin. But the usual dynamic prevailed: one developed nation proposed the species for CITES inclusion and a host of other developing nations along with Japan torpedoed the process, even though everyone knows full well that a complete moratorium is the only thing that would be truly fair to all nations—a whalelike scenario in which the bluefin would be taken off the table for everybody. Some tuna advocates are coming to the conclusion that, as with whales, a different tack has to be taken, one that has more to do with the popular consumer mind-set than with science and policy. One that would ask consumers to evaluate all the negatives of bluefin tuna and end the fish's plight by choosing not to eat it.

The first step in effecting any kind of change for any kind of issue is drawing attention to it. I think with fish we are still in that phase of 'awareness enlightening.'" These are the words of Vikki Spruill, a dedicated ocean conservationist who is one of the key minds behind Seafood Choices Alliance, an organization of organizations that all, in their own way, try to spread the message of eating correctly from the ocean. A slender blond woman in her mid-fifties, Spruill grew up fishing on the Florida coast and had originally started out academically trying to become a marine biologist. But because of what she saw as the male-dominated nature of the discipline, she drifted away from her first love and moved into a career in public relations. She might have continued her corporate work, advising major American businesses on marketing strategy, had the Pew Charitable Trusts not tapped her to help with a public-relations initiative that had been brewing ever since passage of the 1996 Sustainable Fisheries Act.

"My charge was to put ocean conservation on the map," Spruill told me. In the late 1990s, she directed a two-year research investigation focusing on American consumers and fish. The goal was to divine what it was that Americans understood about the sea and what it would take for them to make ocean conservation a higher priority. "What we found is that people relate to fish as much as food on their plate as they do wildlife. People wanted to continue to eat seafood. So we started looking at that 'food/plate connection.' We thought, 'Okay, let's think about how we can build a campaign around getting people to protect seafood because they want to protect the livelihood of the fishermen and the lives of fish.' "

After much deliberation she concluded that it was necessary to focus on a single very big fish to make their point, because "you have to create these visual images for people." Spruill and her team considered the bluefin tuna but in the end felt they needed something with better research and a greater potential for recovery. This led them to another large, open-ocean fish that often crosses paths with the bluefin, the North Atlantic swordfish. "We picked swordfish, first and foremost because if you look at any fisheries graph for swordfish, you just see an absolute straight line plummeting down since the 1960s. We had really good documented science on the status of North Atlantic swordfish. . . . And we liked swordfish because it was something that people were already eating; they were already connected to it."

Beginning in 1998, Spruill's organization, SeaWeb, in conjunction with another U.S. nonprofit called the Natural Resources Defense Council, launched "Give Swordfish a Break"—perhaps the first-ever fish-abstinence campaign that encouraged primarily chefs but also consumers to eschew swordfish on the menu. The campaign launched in 1998 with the endorsement of twenty-seven prominent chefs, then quickly enlisted the support of an additional seven hundred chefs at restaurants around the nation. Participating chefs were required to agree to a Give Swordfish a Break pledge and not to serve the fish in their establishments. Dozens of businesses, including hotel chains, cruise lines, supermarkets, air carriers and others, also removed North Atlantic swordfish from their menus. But there were limitations placed on the program from the outset. As Spruill told me, "We didn't want one of these open-ended, never-finished boycotts." The campaign had a discrete, realizable goal—to close portions of the swordfish's breeding grounds in the Gulf of Mexico to fishing during spawning season. After two years of the campaign, the U.S. National Marine Fisheries Service agreed to close swordfish nurseries to fishing. The campaign was officially ended. Two years later scientists found that there had indeed been a remarkable recovery for swordfish. From a point of near collapse, the population of these majestic ten-foot-long animals had gone from 10 percent to 94 percent of what biologists considered to be their historical population numbers.

The campaign is a success story in every aspect. A clear goal was identified. An appropriate citizen response was crafted. But it's worth looking into what *really* changed the swordfish's fate. In the end it was not a diminishment in actual swordfish consumed that changed the dynamic but rather strong, unilateral government action. There were probably no fewer swordfish caught and eaten during the time of the campaign than before its launch. Rather, it was the *threat* of turning swordfish fishing (and perhaps fishing in general) into a pariah that raised media attention and pressured the fisheries service into closing the swordfish's spawning grounds and protecting the long-term viability of the stock.

But this very effective campaign has had a very murky effect on the public's perception of what is required to save other big, sensitive fish like the bluefin. Ever since the Give Swordfish a Break campaign, more and more nonprofits began embracing the idea of choosing "good" fish over "bad" fish as a means of saving the ocean. Today millions of consumers carry around seafood safety cards with lists of good fish (frequently labled green), sort-of-good fish (yellow), and outright-bad ones (stop sign red). But frequently these lists are not connected to specific policy goals, leading consumers to believe that just through their abstinence they are saving the sea. With globally caught and consumed fish like bluefin tuna, though, one consumer's abstemiousness is nearly always shadowed by another consumer's appetite. And with bluefin, where the ICCAT, the regulating authority of record, seems to hold limited sway with fishing nations, all the negative reviews in the world do not seem to be helping. Bluefin tuna is now on every single red list on every single wildlife-conservation organization's seafood card. But in the time period from when nonprofits began listing bluefin in their various "do not eat" columns, global consumption of bluefin has only increased. U.S. demand has indeed declined, but the Japanese demand has increased to the point where there are no longer enough big wild tuna to fulfill the needs of the market.

In the last three years, the United States could not catch enough fish to meet its legally allotted bluefin quota. In the Mediterranean, where fishermen are also running out of big bluefin to catch, a perverse "farming" strategy has been launched in which wild juvenile tuna are netted, put in pens, fattened, and then sold as "farmed" tuna.

It is important to note that this represents an overall loss of wild bluefin, not a gain. In these ersatz farms, bluefin juveniles are being removed from the wild and denied a chance to breed. And so bluefin are now being eradicated at both ends of their life cycle. The big thousand-pound breeders are being caught and sold as wild fish. Their offspring are being netted in the tens of thousands and fattened for human consumption on "farms." Neither ends up getting a chance to adequately reproduce. Both Eastern and Western stocks of Atlantic bluefin are collapsing.

Not even the threat of mercury poisoning seems to have had any effect on bluefin consumption. Just as in the debate with PCBs and salmon, mercury and tuna have become linked over the last twenty years to the point where most consumers are aware of a tuna/mercury connection. Like PCBs, mercury enters the oceanic food webs when the pollutant is released into coastal ecosystems. The most famous example of coastal mercury contamination occurred during Japan's post–World War II industrial expansion. Throughout the 1940s and '50s, the Chisso Corporation consistently discharged waste products from the manufacture of the industrial chemical acetaldehyde into the enclosed, poorly circulating waters of Minamata Bay, Japan. Thousands of people eating local seafood came to suffer from what came to be called Minamata disease, with extreme birth defects and early mortality experienced throughout the local population.

Today this kind of egregious dumping of mercury is rare, but mercury is still entering the environment on a regular basis from coal-fired power plants throughout the industrialized world. This happens when mercury deposits in coal seams are "methylated"—that is, fixed through combustion to carbon and hydrogen atoms—into a chemically "sticky" molecule called methyl mercury. Methyl mercury bonds readily to living tissue when introduced into the marine environment.

As with PCBs, methyl mercury is first absorbed by plankton and then passed up the food chain to small forage fish, then to low-level predators like mackerel, and then finally on to apex predators like tuna. Again, as with PCBs, methyl mercury has a tendency to linger in animal tissues (though not nearly as long as PCBs). Therefore, as with PCBs, mercury concentrations amplify in fish at higher levels on the food chain. It is the biggest, longest-living fish that tend to have the most mercury, and in the ocean it is harder to find a bigger, longer-living fish than the bluefin tuna. Some U.S. consumers have backed away from eating bluefin. The "choice" of eating unpolluted fish versus polluted fish is another factor often included in safe-seafood lists compiled by American nonprofit organizations. But, curiously, Japan, the place that had the most extreme exposure to mercury poisoning, keeps eating big tuna with abandon.

It starts to feel as if the phase of "awareness enlightening" with big fish that Vikki Spruill spoke of, the phase in which consumers educate and, dare I say, edify themselves by choosing good fish over bad fish, needs to come to a close. For in the end, this somewhat passive response to the global crisis in fisheries robs the conservation movement of the will to stage more radical, directed, and passionate action. Daniel Pauly, the author of the shifting-baselines theory and frequent critic of the limited views of the sustainable seafood movement, said as much in a recent paper. "The current faith in the magic of free-market mechanisms must be questioned," Pauly wrote. "Consumers should not be misled that a system of management or conservation based on purchasing power alone will adequately address the present dilemma facing fisheries globally." Indeed, Pauly's words turned out to be prescient even with the "saved" swordfish. Harpoon-caught North Atlantic swordfish is now listed as a "Best Choice" or "Good Alternative" on the Monterey Bay Aquarium's seafood watch cards. But the National Marine Fisheries Services has relaxed fishing for those same swordfish where they breed in the gulf, and fishermen are now reporting severe declines in the stock. As Adam LaRosa, a charter boat operator who targets tuna and other big pelagic fish, told me, "These kinds of things don't work, unless you do them forever."

Meanwhile the bluefin catch continues, the fish decline. No one has yet motored a Greenpeace Zodiac between a school of breaching bluefin tuna and the boat that would haul them in to market. Whales have become wildlife. But tuna have remained food. The "seafood choices" wing of the ocean-conservation movement would ask people to hold a dual concept of the bluefin as both food and wildlife, but this doesn't seem to be something humans can do. To most people an animal is either food *or* wildlife. If a fish ends up in the market, humans will come to the obvious conclusion that it is food; they will then choose to eat it, even if they are warned that the fish is endangered or contaminated with mercury. In the absence of a larger moral argument and more profound government action, the animal's appearance as flesh in the market, unfortunately, argues more effectively than do any caveats against eating it.

One important element of the whale-conservation movement that the environmental victors seldom discuss is the fact that by the time the Greenpeace Zodiacs hit the waves, whales as objects of commerce had become more or less irrelevant. While Norwegians, Japanese, and Icelanders continued (and continue) to eat a few pounds of whale meat every year, the majority of humans, once their economies had recovered from the devastation of World War II, no longer needed whales. One of the driving forces behind the second era of whale hunting was the production of edible oils from whale fat for margarine and cooking. Throughout the sixties and seventies, though, there was an edible-oil revolution (that Green Revolution again). Developing nations were encouraged to produce palm, ground-nut, peanut, and other agricultural oils locally. By the time of the 1982 moratorium, whale oil provided less than 1 percent of the world's cooking-oil needs.

It is here perhaps that a parallel development with bluefin tuna needs to be pursued. If no one is willing to get into that damn Zodiac, then a replacement for bluefin must be found, something that would undermine a major component of the fish's attractiveness. A domestic version of the fading wild animal.

But as we reach the last chapter of fish decoding, we are finally coming up against a fish that, while we like it a lot, may ultimately not make very much sense to farm. Tuna-ranching operations whereby wild young tuna are netted, transferred alive to pens, and fattened to adult sizes have been in existence for over a decade now; indeed, these operations today remove more tuna from the wild than do traditional fisheries. And now, with tuna ranching facing staunch environmental criticisms, a brave new world of fish domestication is nearing realization, one that owes its existence to the decoding of the European sea bass back in the 1960s and '70s.

As I write this, the final steps of closing the life cycle of bluefin tuna have just occurred. A company called Clean Seas in Australia has created perfectly temperature- and light-controlled breeding tanks that have induced the first captive spawning of bluefins in history. The time-release GnRH hormone spheres developed for sea bass and sea bream by Yonathan Zohar (the Israeli self-described "ob-gyn for fish") are shot into the tuna via a slender harpoon. Thanks to this technology, in July of 2009 the first large-scale captive spawning of tuna took place under a thick cloak of secrecy. *Time* magazine judged the achievement as the second most important invention of the year. Soon after the *Time* article, Zohar wrote me, deeply troubled, that he and all his work that had preceded this milestone got nary a mention in the *Time* article.

But beyond the spawning of tuna, there are considerable complications ahead. Because bluefin are warm-blooded and lightning-fast, they have furiously high metabolic rates. The microdiets of rotifers and artemia used with sea bass aren't enrichable enough to satisfy tuna's high-energy demands. Moreover, maintaining a family of five-hundred-pound tuna broodstock is extremely costly. So costly that some researchers are even exploring a bizarre hybridizing project whereby mature bluefin gonads are implanted in the body of a dramatically smaller fish called a bonito in order to turn these small fish into surrogate mothers. Taxonomically speaking, bonito are tuna, but they are less than ten pounds at maturity. It is a weird kind of package, but initial trials are showing some positive results.

However, whether bluefin tuna will artificially reproduce via bonito or via a precisely controlled artificial environment with Zoharian hormone spheres trickling through their bloodstreams, the fact remains that bluefin are warm-blooded, fast-swimming, highly complicated animals that in the best-case scenario will still require a tremendous amount of food to bring them to market. Whereas twenty generations of selective salmon breeding in Norway have brought the feed-conversion ratio in Atlantic salmon down below three pounds of feed to one pound of salmon, tuna still require as much as twenty pounds of feed for every pound of flesh they produce. This is considerably worse than all other fish. Perhaps a selective-breeding program might bring the conversion rate down around five to one, but this is still terrible.

So we have to ask ourselves, is bluefin tuna really so special that no substitute will do? Japanese defenders of the bluefin trade cite the long cultural tradition of tuna sushi in Japan. But as I wrote earlier, when you look at it in a historical context, the Japanese have a very *short* tradition of eating bluefin. Before the American occupation of Japan, Japanese preferred lean fish and meats and found the bluefin too fatty to stomach. It was only after the American occupation and the subsequent introduction of fatty beef into the Japanese diet that a taste for the fatty "toro" belly of bluefin started to become fashionable. If the Japanese adapted to a higher-fat diet in less than half a century, can we not shift gears again and adapt to a sustainable diet in the same period of time?

It was in answer to these questions that I set out trying to discover a truly thick-fleshed farmed fish that could fulfill the steaky category most seafood diners now expect to see on a menu. A fish that had the "bite" of tuna but might have a footprint more akin to that of a barramundi or a tilapia. And so I found myself in a dive boat, three miles off the coast of the big island of Hawaii, motoring across the cerulean blue of the South Pacific with a tall, highly optimistic Australian named Neil Sims. Rejoicing in telling me tales of his adopted land, Sims was flitting from topic to topic, bearing the relaxed but enthusiastic attitude of what Hawaiians call the "aloha spirit." Eventually we neared the site of Sims's farm—a huge underwater ziggurat that is the center of his company, Kona Blue.

It had been a long time since I'd scuba dived, and even when I'd first learned in college, my skills had been rudimentary at best. We were now about to embark on what is called in scuba a "blue-water dive"—a plunge into the open ocean, hovering over a depth of nearly three hundred feet of water. Because such a dive does not take place above a coral reef, a seawall, or any other structure, a blue-water dive is extremely scary. There are no perceivable reference points that allow the diver to determine depth. If the diver is, like me, inexperienced, he can freak out, lose his bearings, fail to establish neutral buoyancy with his buoyancy-compensator vest (a kind of external human swim bladder), and find himself sinking unstoppably. If this happens, the diver will surely die. Either he will be crushed like a tin can at the bottom or, because he sank so low that he did not have adequate oxygen to surface slowly and decompress, oxygen will boil out of his blood and block his veins and arteries when he dashes to the surface. I was busy trying to keep cool and not betray the fact that I was scared shitless.

This all crumpled when Sims patted me on the back, looked me in the eye, and said, "Got your wet suit on backwards, mate."

In spite of my anxiety, I was curious to see what Sims was up to. He was, like Josh Goldman and his barramundi in Turners Falls, flying in the face of convention when it came to his selection of fish. Up until very recently, most of the fish that we've chosen for our consumption and domestication have been accidents. We have taken those species because we knew them as wild game and then found that they fit well into our culinary and economic niches. We seldom considered their biological profiles or whether they jibed well with conditions that humankind could provide them.

Norway selected Atlantic salmon as its target farmed fish because the demise of wild temperate-zone rivers around the Northern Hemisphere was a common plight. Most Americans or Europeans had a distant memory of wild salmon, but practically no one had access to a reliable supply of it. Farmed salmon reclaimed that lost memory.

Israelis chose to pursue the domestication of the European sea bass because it was known in nearly all countries of the Mediterranean and because it was so overfished that it fetched a high price.

Cod became the first global fish commodity, mainly because it took well to preserving—dried cod lasts for years and could be shipped around the globe even on the slowest of oceangoing vessels. But when farmed, cod is expensive and slow-growing—disastrous as an aquaculture product.

But Sims came to aquaculture through environmental zeal, not with the intention of making a buck. And it was his direct personal experience with the limitations of fisheries management that convinced him that fish farming was a better choice than fish catching.

Sims began his career in the remote Cook Islands of the South Pacific. There he was responsible for establishing a fishery-management regime for a kind of giant snail called a trochus that produces an attractive pearly shell, valuable to jewelry makers. Over the course of five years, he tried to implement a number of different approaches to get the Polynesian natives to conserve the trochus stocks, reminiscent of some of the many measures that have been taken with salmon, bass, cod, and tuna. He closed fishing seasons, planned reserve areas, established size limits, reduced individual quotas. Nothing worked. Finally, after trying numerous approaches, he, as the senior scientist, simply closed the entire fishery. The following day he came across a bare-chested Polynesian elder paddling a dugout canoe through the lagoons. Sims looked inside the hull and saw it filled with trochus snails, in spite of the closed season. "I yelled at him," Sims remembers. "Then he yelled at me. He started to cry. Then I started to cry, and then the old bugger finally says, 'Why? Why did you close the season? There are still some trochus left! We haven't caught them all yet.'"

This led Sims to realize that we needed to do more than just regulate fisheries—we needed to work out a different methodology altogether.

A chain of events led him to Hawaii, but the primary draw was the opportunity to begin aquaculture in the clear, strong currents surrounding the town of Kailua Kona. At first he tried pearl farming, but when the pearl market became flooded with freshwater pearls, Sims began reexamining the possibility of returning to his passion—fish and fisheries biology. There were small government grants available for research into marine aquaculture. "People were trying out the Hawaiian fish called moi. It's a niche species, really. And they were also trying milkfish and mullet." None of these species, Sims felt, truly addressed the niche that needed to be filled by aquaculture—the niche of thick-fleshed predators such as tuna.

It was at this point that, like Josh Goldman and his barramundi, Sims decided to turn the equation of aquaculture on its head. Instead of finding a fish that people knew, that was scarce, and that had an established market, Sims wanted to find a fish that was right for aquaculture, whether or not it was known. To rationally and scientifically apply Galton's principles of domestication and see if there was a fish that fit those criteria.

Eventually Sims came across a fish that previously had no market value whatsoever. *Seriola rivoliana*, known as the Almaco jack or the kahala in Hawaii, is a speedy, firm-fleshed blue-water species of the same family of fish as yellowtail and amberjack. Kahala are only distantly related to tuna and do not have their ruby-red color, but they still have the thick, dense flesh of tuna and could easily pass for white albacore if prepared as sushi.

Another important factor about kahala is that they were never fished commercially and are hence quite abundant. The reason for this is that in their wild form the fish can cause a disease in humans called ciguatera poisoning. Ciguatera is a poison created by microscopic organisms called dinoflagellates. The toxins in dinoflagellates enter the food chain like all the other toxins mentioned earlier—that is, from the bottom. Dinoflagellates stick to coral and are eaten by small fish. These small fish are then eaten by bigger fish. Just like mercury, ciguatera toxins generated from the dinoflagellates bioaccumulate in large predators like kahala making kahala flesh dangerous food for humans.

But when kahala are fed a traditional aquaculture diet and isolated from tropical reefs, the fish are ciguatera-free. And because the wild population of kahala is large and healthy, they are unlikely to be severely damaged through interaction with farmed populations. Moreover, of all marine fish currently farmed, kahala have among the best feed-conversion ratios ever achieved. Without any selective breeding whatsoever, the amount of fish required to produce a pound of kahala ranges from 1.6-to-1 to 2-to-1, ten times better than the feed conversion ratio for bluefin tuna. Feed trials scheduled to begin in summer 2010 will introduce pellets without any wild fish meal at all.

As for another of Galton's principles, the one that stipulates that "they should breed freely," kahala are equally appealing. When I asked Sims later if he uses any of Yonathan Zohar's time-release polymer spheres or photoperiod manipulation to get the fish to spawn, he responded cheekily. "No, we do not use any hormones or environmental manipulation. We tried soft music and candlelight and a little wine, and it worked just as well without. So we kept the wine for ourselves." Kahala spawn constantly, sometimes weekly, throughout the year. They are, in short, the fish we should have chosen right from the start.

The problem is, as with the barramundi of Turners Falls, no one quite knows what they are. Neil Sims and a marketing team that included the main investor in Horizon organic milk have decided to call the fish "Kona Kampachi." Kona for its point of origin and kampachi based on a similar fish that is consumed in Japan. A sushi chef in New York whom I later asked about the fish complained, "Well, you know, Kona Kampachi, that's an artificial name. Kampachi is kampachi, and it is from Japan."

But artificial name or not, the fish has real benefits and poses a real possibility for change. Diving into the waters around Kailua-Kona, watching Neil up ahead of me, I felt the sensation of a whole different world emerging before me. Using technology developed over the last ten years by the University of New Hampshire, Kona Blue has constructed diamond-shaped cages that can be moored in the open ocean. While powerful storms do happen off Kona and a rupture could occur in Sims's nets, the fact that the fish Sims is using are not selectively bred limits the potential genetic impact the fish could have on the surrounding populations should they escape. As I glided down, down, down, past the beautiful fish swimming in unison in their net pen, I thought that for the first time I was seeing the ocean on a fish's terms. The site of these pens had been painstakingly chosen; the swift, swirling currents mean that nutrients do not accumulate below the pens, and therefore the impact on the environment is minimal. Sims also constantly monitors his kahala for ecto-parasites like sea lice and has found their occurrence on his farm lower than on kahala in the wild. Down and down I drifted. From below I looked up at the cage, seeing how little it looked in relation to the bigness of the ocean.

Suddenly I saw a human hand reach over in front of me and grab my diving vest. In the silent communication that happens underwater, I could read the grave concern in Neil Sims's eyes. He looked at me wide-eyed and pointed down. I glanced below and saw the huge, gaping maw of the lifeless ocean beneath me. I had incorrectly set my buoyancy compensator, my human swim bladder, and if he hadn't grabbed me, I was well on my way to sinking into the eight-hundred-foot trench below. Sims expertly inflated my vest. I began to float easily, and my breathing quieted.

Sims waved me over to the side of the net pen. I floated above him silently, close enough to see that the fish actually seemed to recognize him. In what he would later describe to me as the "rock-star effect," the fish crowded to be close to him, expecting from him some kind of deliverance or gift or both. Sims spread his arms out wide and seemed to take in their adulation.

Kona Kampachi has over a 30 percent fat content, higher than most tuna. It retails for eighteen to twenty dollars a pound in fillet form and to date has a tenuous foot in the market. Production reached over a million pounds in 2008, about half the total world catch of bluefin tuna. It does not have the rich ruby color of tuna (a color that is often enhanced artificially by "gassing" tuna with carbon monoxide), but it is an extremely pleasant sushi experience—it satisfies the sashimi yen that has been created over the last twenty years—the yen for the firm, ATP-rich musculature of a fast-swimming pelagic fish.

And for those who would still favor tuna, Neil Sims is quick to point out the essential imbalance between humans and those great fish. "Is tuna farming really going to be able to sate the panting palates all around the planet? We certainly cannot do it on the backs of wild bluefin or wild yellowfin any more than we could sustainably feed the world with wild woolly mammoths."

Kona Kampachi is slowly getting a reputation. It is, like Josh

Goldman's barramundi, like the tilapia and the tra, a good idea. But as the world tries to emerge from financial crisis money for ventures like Kona Blue may dry up. Can we embrace a whole new set of species that we don't know intimately? Can true sustainability rise above the noise of so many pretenders to that name? Can we come to an understanding of which fish work for us and which fish don't? I would hope so. I would hope that these traits, these characteristics, become the traits and characteristics we desire most. Our survival and the survival of the wild ocean may depend on it. I took one more look at Neil Sims floating below me with arms outstretched, his kahala finning in the current each one mutely appraising this conductor of an all too silent concert. The only sound was the whir of bubbles boiling by my ears up toward the silver mirror of the surface above.

I got one last chance to go out tuna fishing before I concluded my research for this book, but this time I was to come on as an observer and not a fisherman—a role I bridled against at first but one that, as the enormity of the problems facing bluefin hit me, seemed more apt. My hosting vessel was a sleek sportfishing boat that travels up and down the East Coast hunting tuna during their annual migration. In summer months the boat pursues the smaller yellowfin and bigeye tuna as a charter operation, but in January she and her crew take up residence in Morehead City, North Carolina, a key stop on the giant bluefin's passage down the coast to its spawning grounds in the Gulf of Mexico. As we set out from port at three-thirty in the morning, I could discern at least a dozen other wakes of the most state-of-the-art fishing vessels, also out for bluefin, piercing the darkness as we sped toward the grounds.

The January bluefin fishery seems a lot like sportfishing, but it is in fact commercial. Once upon a time, bluefin were numerous enough to allow harpooners and netters of different sizes to pursue the fish. But now, with the fish's numbers in severe decline, only hook-and-line fishermen are allowed to attempt to catch them. As the boat's engines stepped down to trolling speed, the mate let out eight different lines rigged with spear-nosed ballyhoo baitfish into the wake behind us. The commercial limit for bluefin was set at two fish per boat, but based on what I had heard, the fleet of a dozen-odd boats would be lucky if they brought in two fish altogether. Still, for the happy hunter who does get one the price is worth it. A single wild bluefin will often sell for more than ten thousand dollars.

Our crew was slightly different from the others out fishing bluefin that day, in that we had with us a cameraman and a professional sports angler from a popular cable-television show called *Quest for the One*. So popular had *Quest for the One* become that the producers were branching out with a sequel, a show called *Monster Fish*. The professional angler, a man in late middle age, was supposed to fight a monster bluefin in what could be a multihour battle.

It was a particularly rough weather day, and the crew was being constantly thrown about as they readied for the fishing. Like picadors and banderilleros in a bullfight, the captain and mate rushed around, prepared baits, guided lines up the outriggers, and generally did all they could to secure the possibility of a strike. The professional angler lay like a matador on the couch in the boat's luxurious stateroom, awaiting the moment when he would be summoned to take on his much larger challenger.

Two fishless hours passed for the entire fleet. Feeling a little queasy, I wandered out to the cockpit to get some air while the mate rigged ballyhoo fish in a manner that he didn't want me to describe or the camera man to film. So scarce were the bluefin in the winter of 2009 and so numerous were the boats that even the small advantage of a uniquely hooked baitfish could mean the difference between a ten-thousand-dollar day and a zero-dollar day. As the mate rigged baits, our conversation fell to the situation across the Atlantic, where bluefin were being either relentlessly hunted for direct sale as wild fish or scooped up as juveniles and sold to tuna ranches.

"It's a sad situation," the mate said, popping out the ballyhoo's eyes and running a wire around its bill. "They're just killing them over there in Europe. I mean, we'd shut our fishery down in a second if they'd stop." I thought of Neil Sims's Polynesian, the bare-chested pursuer of trochus snails who cried when he was told that the trochus season would be closed. "But there are still some left!" the old man had said. Was there really so much difference between that old man in the dugout canoe and this college-educated American in the sleek fiberglass hull of a sonar-equipped, half-million-dollar sport-fishing cruiser? The owner of the boat later made the very accurate assertion that the purse seiners catching juvenile bluefin in the Mediterranean are catching a hundred times more fish than his boat trolling off Morehead City. His boat was allowed to catch only ten bluefin a year, and the boat stayed very much within its legal limit. But still, those ten bluefin are some of the last huge breeders that play a critical role in the survival of the stock. The Morehead City boat was following the rules. Many fishermen in the Mediterranean are not. But everyone is still fishing. No one is stopping. These thoughts filled my head as I returned to the cabin and nodded off to sleep.

After two more hours of dragging baits in the penumbra, the mate poked his head in the door.

"We got something up ahead," he said excitedly. The professional angler rose with what seemed to me to be a slight whiff of boredom. Another long fight with another big fish. I, on the other hand, who had never seen a giant bluefin, bounded up the ladder that led to the console where the captain scanned the horizon eagerly.

"Look at that shit up ahead," the captain said. "It's fucking raining birds."

I'm slightly nearsighted, and at first I couldn't make out what he was pointing to, but as the boat moved forward, I saw seabirds gathered up into a cloud, the size and violence of which I had never seen before. Gannets—big, albatross-like pelagic birds—flew hundreds of feet above the churning surface of the water. In a flock of many thousands, they whirled in unison and then, as if on command from some brigadier general of bird life, dropped in an arc, bird after bird, into the water beneath. The gyre of gannets turned in a clockwise direction, and down below, spinning counterclockwise, was the largest school of dolphins I'd ever seen. There in the angry blue-green sea, the dolphins had corralled a vast school of menhaden—small herringlike creatures that, when bitten, release globules of oil that float to the surface. Oil slicks flattened the water everywhere as the dolphins swirled around, using their exceptional intelligence and wolf-pack cooperation to befuddle and surround the fish, which in turn whirled in a clockwise direction.

It was one of those rare moments where one has a vision of the scope of the wild ocean. Not just small cylinders firing to keep a tiny engine running, but rather the giant, massive gears of nature, each one with its own reasoning, its own meta-logic, spinning in its particular circle in competition or in confluence with the gear below it. We zeroed in on the school, but our progress was painfully slow.

It would have been foolish to speed into the midst of the tumult—we would have ruined our baits in the process and doomed our chances of hooking a tuna.

But, luckily, the commotion did not subside. If anything it only grew more frantic and exuberant on our approach. Beneath the birds, beneath the dolphins, beneath the menhaden, there should have been an equally vast school of giant bluefin tuna, collaborating with vertebrates of the so-called higher orders of life to form the floor of the prey trap, sealing the baitfish in from below, while the dolphins and birds made up the trap's walls and ceiling. A strike from a giant tuna seemed inevitable. The professional angler cracked his knuckles below in the cockpit. The mate scanned the outriggers.

But as we passed through the orgy, it appeared that this trap had no floor. Only dolphins, an animal humanity has decided are "good" and worthy of preservation, breached endlessly in the white water around us. Only gannets, another animal that has similarly been deemed "wildlife" and is no longer shot and killed, swirled above us and plunged like a global squadron of dive bombers into the sea below. The vast machinery of the food web spun out before me and would continue to spin, conceivably for millennia to come, with our tacit approval. But the final gear in the system, the tuna, the part that interested me most, was missing.

Those who study fish or pursue fish or live among fishermen love fish dearly. Meanwhile, the rest of the world eats more and more of them every year without ever really bothering to learn what any of those fish look like, how they behave, or how many remain. I hold on to the hope that the dynamic might change. That fish might one day be understood as their own kind of perfection, meriting their own special kind of respect. Recently I asked a biologist who had spent his life studying tuna whether he thought that bluefin could ever be elevated to the status of a whale or a dolphin and given protection akin to that afforded the other great animals on earth.

"What I always say," he told me, "is that in the early days of the founding of the United States, right there in the Constitution it said that a black man was once worth three-fifths of a white man. And look at us now. Never say never."

Conclusion

Whenever I told people that I was writing a book about the future of fish, I would typically get two reactions. The first was the urbane, witty response. "Oh?" my interlocutor would say. "I didn't know fish had a future." Though it was flip and shortsighted, I didn't mind this reply. People generally don't like to look an ugly and serious problem in the eye, and the redirection implicit in this comment was, in a way, very honest and very human.

It was the second response that I found more troubling.

"Oh, you're writing a book about fish. Which fish should I eat?"

Perhaps it is a particularly American trait—the belief that the individual by his or her personal actions can somehow shift the course of history. But when it comes to choosing the "right" fish, the sentiment I first noticed in the United States has spread to other nations, to the point where a veritable chorus rises up from any table I visit, be it in England, France, South America, or Asia, every time I mention my damn fish book.

"Which fish should I eat?"

Choosing a fish that is well managed or grown on a farm that uses sound husbandry practices is most definitely personally satisfying. One feels "good" when one eats "well." It is not for nothing that the Buddha himself included sound eating practices as part of the path to enlightenment. "Do no harm," the Buddha spoke, "practice restraint according to the fundamental precepts, be moderate in eating. . . ."

But the public's choosing of "good" fish in the marketplace has had little effect on the actual management of wild fish or the practices of growing farmed ones. The Monterey Bay Aquarium—which has distributed over a million seafood cards that label fish as "red" (avoid), "yellow" (good alternative), and "green" (best choice)—took the brave act of commissioning a survey of the programs' effects. The results were telling: fishing pressure had not been significantly reduced on any of the species or stocks consumers were advised to avoid.

In defense of the Monterey Bay Aquarium, I don't believe that the program's innovators thought seafood-advisory cards would actually change fish-consumption patterns. First and foremost, the ratings cards were conceived of as tools for public education. Prior to their introduction, relatively few people knew about the overfishing of bluefin tuna, the negative effects of farming Atlantic salmon, or even the existence of good fishing practices and bad ones. People generally saw individual species the way Mark Kurlansky's mother saw cod: "fish." A crop, harvested from the sea that magically grew itself back every year. A crop that never required planting.

The historical vocabulary around fish echoes this sentiment. Think of the word "seafood" itself. How many genera and species are described by these two opaque syllables? Equivalents in other cultures are no less vague or misleading. In German, French, Spanish, and most of the other Western European languages, seafood is "sea fruit." Slavs, meanwhile, often call the many creatures of the oceans "gifts of the sea." All these expressions imply that the ocean's denizens are vegetative, arbitrary, and free of charge. So-called vegetarians, indignant over the suffering of farmed cows and chickens, frequently include wild fish in their diets. Kosher laws that mandate the merciful slaughter of mammals and birds do not apply to fish.

Thanks to the Monterey Bay Aquarium and other organizations, we are now at a point where we know something about fish. We know that overfishing can and does happen. That, as with terrestrial animal husbandry, fish farming has problems of waste management, disease, and industrial pollutants. We are not Neolithic cave dwellers, showering this flock of passenger pigeons with arrows or driving that herd of mastadons over a cliff. We have inklings of what it is we are doing.

Nevertheless, we are still not grappling with the quandaries of fishing and fish farming in a manner commensurate with the contemporary battles of the food-reform and land-based environmental movements. We are now a bit like the jury in the 1818 *Maurice v. Judd* case. Whereas that jury sequestered itself to decide whether or not whales were fish, we are now deliberating over whether fish are wildlife—wildlife that is sensitive to our actions and merit our sound protection and propagation.

It is not that we don't have choices to make. But the choices ahead are large societal ones that require our careful attention and our active political engagement. After forty years, beginning with the near global collapse of wild salmon, to the revival of the American striped bass, up through the closure of the cod fisheries on the Georges Bank and Grand Banks, and on into the rise of more sustainable-aquaculture alternatives like tilapia, we have seen numerous examples of oceanic disasters interspersed here and there with real improvements. Wild fish globally are declining, but the examples of science-based successes are marked, accurately documented, and clearly replicable. Pollution and dead zones have grown, but, unlike the terrestrial environment, the essential habitat of much of the world's marine life remains reclaimable. On dry land, urban sprawl consumes 2.2 million rural acres a year in the United States alone, but there is no equivalent development of the sea. If left alone, marine ecosystems have a tendency to rebuild themselves. Global warming is changing oceanic conditions, but fish have survived extreme climate change before and can again. Although ocean acidification is a real and growing threat, a rebuilt and robust wild fish population could help buffer ocean pH. Fish excretions, it turns out, are on the basic side of the pH spectrum. A radical increase in wild fish could be a bulwark against acidification.

What is needed now is a societal choice to give priority to a set of clearly achievable goals for wild fish. Those priorities should include:

1. **A profound reduction in fishing effort.** The world fishing fleet is estimated by the United Nations to be twice as large as the oceans can support. This overcapacity is being maintained primarily through government subsidies. Many billions of dollars are paid by governments to support fishing fleets that without subsidies would not turn a profit. Subsidies thus make wild fish unreasonably cheap. A move away from large, heavily extractive (and heavily subsidized)

vessels that employ very few individuals is critical. An emerging "artisanal" sector of respectful fishermen-herders that will steward the species, as well as catch them, needs to be encouraged and higher market prices will be able to support that kind of activity.

2. **The conversion of significant portions of ocean ecosystems to no-catch areas.** Up until the last decade, the default assumption with the ocean has been that any ocean habitat could and should become fishing grounds if fish are present in abundant numbers. There is, however, growing evidence suggesting that key fish breeding grounds and nursery habitat must be reserved as safe havens if overexploited fish populations are to rebuild to harvestable numbers. It is still a matter of controversy how much territory should be put aside for fish reserves, and today an average of only 1 percent of the world's ocean habitats is protected from exploitation. Surely developed nations that already protect around 10 percent of their land areas could see fit to come up with a similar amount for their ocean holdings. Rather than eating into our principal as we've done for the last thousand years, by setting up a network of fisheries reserves we will in a sense put a portion of our ocean wealth into low-interest municipal bonds, an investment that if left alone will pay a steady, compounded interest over time.

3. **The global protection of unmanageable species.** Species or stocks that straddle too many nations or that occur in unowned, international waters have been shown with very few exceptions to be unmanageable over the long term. In the face of hard science, politicians of multiparty treaties "negotiate" catch allocations that go against scientific reality. Developing nations balk at not being given their "fair share" of these depleted stocks, but if a species shows continued decline over time, as has the Atlantic bluefin tuna, the

only "fair" thing to do is to completely close the fishery. In some cases it may be advisable to consider certain species simply too valuable to hunt. If bluefin tuna were elevated and accorded the same kind of protection tigers, lions, whales, and other sensitive transboundary species are given, it could shift public perception of fish and give regulators a line in the sand past which a species is simply not allowed to decline.

4. **The protection of the bottom of the food chain.** With the boom of aquaculture and the rise in the use of fish as feed for pigs and chickens, small forage fish like anchovies, sardines, capelin, and herring now represent the largest portion of fish caught. All of these fish are in greater and greater numbers being ground up in reduction facilities and recast as food for fish farms and terrestrial farming operations. And yet we really do not understand the population dynamics of these smaller forage fish, and we do not really know how to manage them. With the scaling-up of so much aquaculture, we run the very real risk of what Dr. Ellen Pikitch of the Pew Oceans Commission called "pulling the rug out from underneath marine ecosystems"—that is, removing the basic food source of the ocean and causing fisheries collapses from below.

According to Pikitch, ecosystem models of forage/predator systems are increasingly showing that intact wild systems are more valuable in raw dollars than are systems converted to aquaculture. When small forage fish are left unharvested, the resulting catch of bigger commercial species that eat those forage animals is greater. There is simply more food in the water, more energy in the system, and that energy is translated into more and bigger fish.

We must therefore take a precautionary approach to the very bottom of the oceanic food chain and exploit those animals only after models have been developed that indicate the extent of removals that the system will tolerate. We must also seek to rebuild the bottom of the food chain we have already lost by restoring the habitats where forage fish are born and reared. Estuaries and river systems are vital zones of *food* production and not simply "natural" spaces. Nearly every wild fish highlighted in this book—striped bass, European sea bass, cod, Alaska pollock, Atlantic and Pacific salmon, bluefin tuna—depends upon a supply of forage fish whose life cycles are in turn dependent upon rivers and estuaries. Herring, menhaden, smelt—all these small fish are the silver coin, the coin of the marine realm, and their hatching and rearing often occur in direct association with access to rivers that enter the sea. Restoring these areas increases the food supply for the fish we eat most. Deny the restoration and no matter how much conservation occurs at sea, abundance will inevitably be limited by a low ceiling of limited food.

Four very good, noble, and ultimately effective principles that will rebuild the seas. Goals that are more and more becoming part of a new phenomenon taking root in conservation policy, that of "ocean zoning." As more users compete for space in the ocean, some places in the world (the island of Asinara off Sardinia and the state of Massachusetts, for example) have implemented overall zoning goals, much in the same way municipalities plan a town with commercial space, green space, and residential areas. The advantage of zoning the ocean now is that it gives wild-fish advocates a chance to stake out territory *before* wildness has been relegated too far to the margins. Hand in hand with ocean zoning is the rising trend of "ecosystem management." Rather than managing individual species, ecosystem management seeks to manage entire systems, modeling patterns for fishing and restoration that work toward reestablishing the balance of the many demands of prey and predator.

But all the very good and noble goals of ocean zoning and

ecosystem management become meaningless in the presence of one ominous factor: human demand.

In spite of campaigns, boycotts, publications, documentaries, and every other means of persuasion known, the global human population keeps growing and humans keep eating more fish every year, not just in aggregate but on a per capita basis. Even with so many warnings about mercury and PCBs, the world nearly doubled its per-person fish consumption in the last half century, from twenty pounds per year in the 1960s to thirty-six pounds in 2005. And because seafood is such a global, boundary-free business, whenever a restaurant, a city, or a country takes to the moral high ground and tries to reduce or improve the footprint of its seafood consumption, another, less scrupulous restaurant, city, or nation is ready to step in and continue the bad practices that the more evolved parties have abandoned.

So if we take as a given that humankind will keep eating fish, more and more of it every year, then we need to come up with a way to direct that appetite away from sensitive, unmanageable wildlife and usher it toward sustainable, productive domesticated fish. A small-scale, artisanal, wild-fish fishery would be a great thing that could inevitably lead to better protection of wild fish. But a small-scale artisanal fishery will never have the industrial capacity of the supertrawlers that decimated the Georges Bank and Grand Banks codfish stocks.

What is needed above all is a standard for boosting fish supplies in as sustainable a manner as possible. Humans should purposefully select a handful of fish species that can stand up to industrial-size husbandry with the goal of compensating for the huge gap between wild supply and growing human demand. Of course, if the global human population continues to grow unabated, no solution will work; in such a population-growth scenario, only the stars can save us. Indeed, with terrestrial food production now reaching its limits, the ocean is, in a sense, the final option, the only remaining way for humans to convert more of the world's biomass and sun energy into more humans. The future of human growth depends largely on how we manage our ocean.

We therefore have a very clear choice. We can carefully select the fish that work well both in conjunction with human farmers *and* alongside the wild ocean food systems that still function. Or we can run roughshod over the wild ocean, install feedlots up and down the world's coasts, and continue to reap short-term calorie credits irrespective of the long-term ecological debits. If humans are at root rational creatures, then we must without question choose the former path over the latter.

It makes sense therefore to return to and expand upon postulates of an earlier era, to revisit the precepts that Francis Galton posed at the dawn of the industrialization of terrestrial animal husbandry. Galton spoke of wild animals outside the dominion of humankind as "doomed to be gradually destroyed off the face of the Earth as useless consumers of cultivated produce." But with the ocean we need both the undomesticated and the domesticated sides of fish to carry forward. It seems, then, that a new set of principles for the ocean has to be made, one *inclusive* of wild systems, systems as nourishing as they are mysterious. We cannot make up for the elimination of our wild-food calories with farmed replacements. We need both—for our nutritional as well as our emotional well-being.

For too long it has been entrepreneurs who have decided which species to domesticate and which to leave wild. Their decisions have been based on market principles and profit, and they have historically not consulted with the managers and biologists who study wild-fish dynamics. This is senseless. If we continue along this pathway, we will only destroy one food system and replace it with another, inferior one, just as we have already done in most of the world's freshwater lakes and rivers. We therefore need a set of principles that guide us forward with domestication, one that is inclusive of impacts on wild oceans. I would propose that the next animals from the sea we domesticate should be:

1. Efficient. In an increasingly stretched world of food resources, we cannot afford fish that require more feed to produce a pound of edible flesh than do our most efficient terrestrial animals. Fish, by their very nature, *should* be more efficient than land animals. Fish do not have to warm their bodies, and they do not have to stand against gravity. All that energy that is wasted in mammals and birds could and should be redirected into growing fish flesh. Thus the warm-blooded bluefin tuna, whose current feed-conversion ratio can exceed twenty to one, should be abandoned as mass-scale farm animals. If a fish like a Kona Kampachi can be produced with similar fresh density at a fraction of that feed conversion, why pursue the tuna?

2. Nondestructive to a wild system. With salmon there is ample evidence to suggest that the culture of farmed variants in close proximity to wild strains can negatively affect wild populations over time. Indeed, if one compares the fate of Atlantic salmon with that of the American striped bass, two fish that were dangerously reduced in the wild and then domesticated, it is instructive to compare their respective fates. Wild salmon populations have generally declined in Maine, Atlantic Canada, and Europe in areas where they interact with farmed salmon. American striped bass, meanwhile, have staged a strong recovery in the wild even in the presence of an aquaculture program that now accounts for 60 percent of all striped bass consumed. The difference? The fish called "farmed striped bass" is a sterile hybrid created by crossing a female striped bass with a male of a related freshwater species called white bass. The farmed hybrid striped bass cannot interbreed with the wild population of striped bass and thus cannot spread its genes beyond the farm. Furthermore, the hybrid striped bass is grown exclusively in freshwater ponds away from the migration lanes of wild striped bass. Wild populations are thus buffered against contracting farm-born diseases.

If the same separation of wild and farmed fish took place with salmon, the remaining populations of wild salmon might do better. Critics argue that the cost of putting salmon in a closed, recirculating system like that used for Josh Goldman's barramundi would make farmed salmon simply too expensive for the average consumer to afford. This is the logical place for subsidies. If we must subsidize fish consumption, then it certainly makes sense to subsidize those practices we know will contribute to a net gain of fish in the world, not cause the destruction of wild stocks. In certain cases—like Chile, for example—where wild salmon are not endemic and there is no obvious impact upon an existing native population, perhaps open-cage salmon farming could still be allowed. But even there the problems of sea lice and infectious salmon anemia are very clear environmental signs that fish should not be farmed too densely and without careful siting procedures. Less-dense stocking of fish farms will cause price increases, but again, this is where subsidies could help to level the playing field.

3. Limited in number. After the technological breakthroughs on feed, reproduction, and husbandry techniques of the 1970s and '80s, it is now theoretically possible to tame pretty much any fish in the sea. In light of this, we have to be on guard against a certain kind of "Gee whiz, I can do it!" behavior among aquaculture researchers. Just because we *can* tame a fish doesn't mean we *should*. Every time a species is brought into culture, new diseases specific to that species or sometimes that genus can develop and spread to related wild populations. Furthermore, new hurdles present themselves with every new species, and a tremendous amount of time and energy is wasted in the early phases of domestication. Instead of constantly trying to bring new species into an imperfect culture merely because we can, we should instead choose just a handful of animals whose rearing we can perfect. Why farm cod when tilapia is already doing the job? Subtle differences in flesh texture, taste, and nutritional content are controllable through feed and rearing techniques and do not require the taming of a new species. If we want variety of species for niche markets, let that variety be provided by small-scale, sustainable wild fisheries.

4. Adaptable. In the debate on aquaculture, environmentalists have frequently taken the position that we should not be farming carnivorous fish, because their overall footprint is larger than that of mostly herbivorous fish like carp and tilapia. Two and sometimes more trophic levels of food consumption have to take place before a salmon gains nutrition from a sardine. Point taken. But the same argument has been made before, most notably with vegetarianism. For many decades now, environmentalists have argued that if all humans were vegetarian, humanity would have a fraction of its current footprint on the globe. I have tried vegetarianism, inspired by this irrefutable logic. And yet I have drifted back to carnivorism, as have many before me. Rather than hoping to change the world by changing consumption patterns, regulations and farm-level reforms need to be put into effect so that unsustainable food doesn't reach the market in the first place. And where better to start this process than with the world's most commonly farmed fish, salmon? On the downside, it seems unlikely that the environmental community will succeed in dislodging the salmon industry from its dominant position in the farmed-seafood sector. On the upside, salmon do seem to have an adaptability to alternative feeds. Seaweeds and soy are increasingly forming the basis of salmon diets and could replace fish meal altogether in the not-too-distant future. As of this writing, at least one company has developed a completely algae-based feed that replaces the need for fish oil and meal in the diets of salmonids. The problem? Once again, cost. Here is another place for subsidies to play a positive role. Let governments make up for the difference in price between wild-fish meal and synthetic algal feeds until the industry has scaled up. It is an investment in the future.

5. Functional in a polyculture. If there is one lesson that has been learned from terrestrial agriculture, it is that monocultures of crops are susceptible to disease and can cause undue environmental degradation. Rather than starting from zero and redoing all of terrestrial agriculture's mistakes, we should start from a place of polyculture, where wastes are recycled as much as possible, where space is maximized for the growing of food, and where *systems* instead of individual species are mastered.

Five principles, then, to lead us to our selection of domesticated animals from the sea. The animals that could and should rightfully be called our "sea food."

As to what we should call wild fish in the future, I leave that to the marketers of what I hope will someday be a more informed and thoughtful fishing industry. But I would suggest that if we continue to eat wild fish, we need to find a new way of identifying them in the marketplace. A set of terms that implies an understanding of fish as wildlife first and as food second. Wild fish did not come into this world just to be our food. They came into this world to pursue their own individual destinies. If we hunt them and eat them, we must hunt them with care and eat them with the fullness of our appreciation. We must come to understand that eating the last wild food is, above all, a privilege.

Epilogue

This summer, for the first time since my childhood, I fished my home waters again—the rocky, quiet bays of western Long Island Sound. I was joined by my daughter, Tanya. As she enters her teenage years, she has emerged as remarkably similar to my mother, in looks if not in character. Blond and slim, she has my mother's dark, bold eyebrows and easy gait. But where my mother was vague, Tanya is concrete. Where my mother was, in a way, helpless and careless, Tanya is directed and given to problem solving. Also, unlike my mother, who fished with me out of obligation, Tanya genuinely likes fishing. I chalk this up as one of the small victories of influence a father can have over his children once they've started down the course toward adulthood.

We were aboard a party fishing boat called the *Angler* out of Port Washington, New York, just twenty miles from New York

City. The *Angler*. I couldn't help but think of Izaak Walton and his book *The Compleat Angler*, the treatise he penned in 1653, that first parsing of fish for the general public. One where salmon were identified as "kings" and pike called "tyrants." I wondered what Walton would make of our intended quarry this night—a specifically American fish called the bluefish, which is so ferocious it would make both king and tyrant flee in terror if they were to come across one on a dark night.

Though the *Angler* was a party boat, capable of carrying fifty or more passengers, this evening only four of us had shown up. It has been a slow few years for party boats. As children increasingly turn away from fishing and toward more contained and less wild forms of entertainment, the old boats of my youth are dropping away. One captain earlier in the summer had told me he couldn't even get enough fares to sail on Father's Day. For a little while, it was debatable whether the captain of the *Angler* would even make the trip this night—the cost of gas alone would guarantee the voyage to be a money loser. But in the end, perhaps because of my daughter's enthusiasm or the boat's newness to the waters (the owners had recently moved from Brooklyn in hopes of pursuing a wider variety of species), the captain decided to sail as a losing proposition.

The other two people on the boat were men, widowers probably, in their seventies. Lonely, windblown characters whose isolation seemed mitigated only by the occasional fishing trip. As we steamed toward the grounds, one of them barraged me with round after round of fishing stories. How on one trip out to Nova Scotia he caught five bluefin tuna up to seven hundred pounds. How on another trip out of Rhode Island he had filled his gunny sacks with nearly a thousand pounds of cod fillets. I calculated all his many fishing trips in my head to come up with a total poundage of animal flesh taken. At least ten thousand pounds of fish must have perished at his hands, even though the average human needs only about four thousand pounds for his entire life. He interrupted me in the middle of my calculations.

"Now the fishing's no good," he told me. "There's nothing."

But within a few minutes of anchoring up and starting a chum slick, the same old man who had decried the nothingness in the waters beneath him saw his pole bend double with the ferocious opening gambit of a three-foot-long bluefish. Dogged and yellow-eyed, it fought hard until the mate stuck it with a killing stroke of the gaff. It landed on the deck with a thud.

There is still a remarkable amount of diverse life in Long Island Sound. In fact, after catching one bluefish myself, I felt an unusual tug on my bait and then the telltale first run of something uncommon yet familiar. Within a few minutes, the mate helped me net a ten-pound striped bass. The first striped bass I had ever seen in Long Island Sound. A fish that was nearly absent from these waters in my youth. Other things that weren't in the sound when I was a child also made an appearance this year. Seals abounded throughout the winter, and in May a school of several hundred dolphins came up the East River and poured into the sound's waters, the first time such an event had occurred in fifty years. The captain of the *Angler* believed that this was because the dolphins had run out of food offshore, that the huge schools of menhaden they traditionally rely upon were being reduced to make omega-3 dietary supplements and salmon feed, leaving the dolphins to starve. But it also could be that dolphins have made a tremendous recovery themselves as laws on marine mammals have led to their greater protection. It's too early to say.

What can be said is that the ocean is both mysteriously dynamic and vastly productive. Never write off the wild ocean. It can always surprise you.

By the middle of the trip, I was happy to see my daughter finally connecting with a bluefish of her own and cranking it in with exasperation. She had never hooked a fish this powerful. When it came up to the side of the boat, her eyes widened with amazement. "Can we keep it?" she asked. I thought about it. Bluefish are, in a way, a perfect symbol of wildness. They are good to eat if they are eaten immediately—they degrade and putrefy within a day or two. They would never make a good "industrial fish." I figured that between smoking them and giving some away, a total of three or four fish would be the right amount to take home, even though our legally allotted limit would be twenty. I gave my assent, and the mate gaffed the fish and threw it into the box.

Just as he did this, the old man who had decried the lack of fish sidled up to me.

"You know, you can take mine home, too. I don't eat fish."

The captain and I stared at him, dumbfounded. This old man who had complained of there being "no more fish" had already killed at least six large bluefish and seemed intent on killing more. To what end?

In former times the captain of a party boat would have shrugged his shoulders and allowed the killing to continue unabated. It is the pleasure of his fares and their regular business that is his quarry. But this captain, a man of my age who moonlighted as a high-school science teacher, was well aware of the peril facing fish and the need for restraint, care, and rationality.

"Okay, folks," he said, "we're not killing any more fish." The gaffs were put away and nets were brought down from the decks above. Moments later my daughter hooked into the largest bluefish of the night. Back and forth it ran, dragging her up and down the starboard side of the boat, around the stern, and then up the port side. As she strained to crank the handle of reel, I felt I knew her thoughts: "I want to possess this thing of great power and beauty. To make it mine and to become its master." Those would have been my thoughts thirty years ago were I standing in her shoes. But as the net came down and drew the fish in, she moved back from the rail. I saw in her the bud of rationality and the logic of both pursuing and saving fish. To know the power of wildness intimately and, at the same time, to recognize the right of that wildness to continue.

The mate brought the fish over the side in the net and then ran to attend to another customer. I carefully drew the bluefish out and kept my fingers away from its snapping jaws. Holding it by the back of the head, I felt the raw power in its muscles—unwieldy, wild, dangerous. I worked the hook from the corner of its mouth and then in one fluid motion put the fish back over the rail. It darted at lightning speed out of sight. The only reminder of it was a wake of phosphorescence as it bounced against some jellyfish and slipped into the depths.

"Will it live?" my daughter asked me, staring into the sea below.

"Yes," I said, nodding slowly. "Yes, I think it will."

ACKNOWLEDGMENTS

In a book of this nature, an author's interviewees are as much teachers as they are sources of quotable information. For the instruction I received, I would like to offer up my thanks to the biologists, ecologists, fishermen/women, writers, and aquaculturists who opened their doors and offered their expertise during the course of my research. Particular thanks for my frequent impositions are extended to Carl Safina, Ted Ames, Daniel Pauly, Mark Kurlansky, Orri Vigfússon, Jon Rowley, Yonathan Zohar, Kwik'pak Fisheries and the Yupik Nation, Tara Duffy, Jeremy Brown, Adam LaRosa and Canyon Runner Sportfishing, D. Graham Burnett, Steve Gephard, Matt Steinglass, Neil Sims, Trevor Corson, Vikki Spruill and the organizers of Sea Web's annual Seafood Summit, the Helenic Centre for Marine Research, the University of British Columbia Fisheries Centre, the World Wildlife Fund, and the Monterey Bay Aquarium. Also most appreciated: the material and structural support provided by the W. K. Kellogg Foundation and the Food and Society Policy Fellowship program, the National Endowment for the Arts Literature Fellowship Program, and the Bogliasco Foundation's Liguria Study Center. In the editorial realm, I extend my thanks to the early publishers of my fish writing—first to Tim Coleman, who ran my first stories in the New England edition of *The Fisherman* back when I was fifteen years old; and later Alexander Star, Gerald Marzorati, Jennifer Schuessler, Amanda Hesser, and Carmel McCoubrey at the *New York Times*. For the editing of the present work, much appreciation and admiration go out to Ann Godoff, Jane Fleming, and Helen Conford at The Penguin Press. The unofficial editorial advice of Cressida Leyshon, Sean Wilsey, John Donohue, David Gold, and David "Mas" Masumoto was also quite helpful. My research assistant, Kayla Montanye, deserves special mention for her hard work on what I hope will be the first step in a long career in environmental writing. Warm thanks to my fellow fisherman, friend, and agent, David McCormick, who saw the literary possibilities in a fish book long before anyone else, and to my many friends and colleagues who weighed in on the manuscript as it advanced. And of course thanks to my mother, who managed to get me a boat when she was broke, and to my father, who took me fishing when he had no time.

Lastly, much love and thanks to someone who provided the continents of emotional support necessary for a man diving into oceans. Thank you, Esther, for your love and patience.

N O T E S

xi   **"Fish is the only grub":** The quotation is attributed to one "Hugh G. Flood," a composite character created by the *New Yorker* writer Joseph Mitchell in a series of articles he wrote about the Fulton Fish Market, later collected as the semifictional account *Old Mr. Flood.* The Flood stories were subsequently anthologized in Joseph Mitchell, *Up in the Old Hotel* (New York: Pantheon Books, 1992).

INTRODUCTION

11   **principal meats:** My summaries of animal breeding and the histories of domestication derive from Trygve Gjedrem, *Selection and Breeding Programs in Aquaculture* (New York: Springer, 2005).

12   **"It would appear that every wild animal":** Francis Galton as cited in Juliet Clutton-Brock, *A Natural History of Domesticated Mammals* (New York: Cambridge University Press, 1999). In addition to his writing on eugenics, animal domestication, and many other topics, Galton was a cousin of Charles Darwin and is considered to be one of the founders of the school of statistical genetics.

12   **around 90 million tons:** Most of my larger macro-level fisheries data are drawn from the United Nations Food and Agriculture Organization's latest biennial report *The State of World Fisheries and Aquaculture 2008*, ed. J.-F. Pulvenis de Séligny, A. Gumy, and R. Grainger (Rome: FAO, 2009), http://www.fao.org/docrep/011/i0250e/i0250e00.htm. The marine ecologist Daniel Pauly and others have repeatedly stressed that the Republic of China's overestimation of aquaculture production and wild catch could significantly skew the overall global data in FAO's statistics. In particular, Pauly takes issue with the assessment that aquaculture is now 50 percent of the world's seafood supply and warns that the actual number may be much lower. While I agree that the data may be skewed, the trend of the rise of aquaculture is unmistakable. If we have not reached a point of 50% aquacultured seafood by now we surely will reach that number within a decade or two.

12   **if history were written by fish:** The observation that World War II represented a reprieve for groundfish in the North Atlantic is based on an interview conducted with Daniel Pauly in the summer of 2005. Other researchers, most notably Jeff Hutchinson

at Dalhousie University, disagree on this point. Whether or not a difference in groundfish numbers before and after World War II can be quantified, it is nevertheless undeniable that fishing pressure declined during the war and that fishing pressure, globally, increased progressively from 1950 through the present day.

13   **"natural selection favors the forces of psychological denial":** Garrett Hardin's oft-quoted essay, "The Tragedy of the Commons," was first published in *Science*, vol. 162, no. 3859 (Dec. 13, 1968), p. 1243, and can be easily accessed at http://www.garretthardinsociety .org/articles/art_tragedy_of_the_commons.html.

SALMON

16   **as many as 100 million Connecticut River salmon larvae:** Data on Connecticut River salmon estimates come from Steve Gephard, director of the State of Connecticut DEP Inland Fisheries Division's Diadromous Fish Program. Salmon restoration on the Connecticut River has at times elicited critical comments, and some have asserted that salmon were never particularly abundant in the Connecticut River because the river's mouth is rather far south in comparison to Atlantic salmon's typical range. Gephard maintains that a large portion of the river's upper reaches are well within the more comfortable latitudes for Atlantic salmon. "We have generated population estimates based on the amount of habitat available to the species (pre–European Contact)," Gephard wrote me in 2009, "and then used production and return rates from the scientific literature to develop estimates on how many salmon that habitat would produce. We came up with 40,000 adults annually. This also seems to jibe with estimates of other rivers in the region. We can never prove it but the salmon biologists in the area seem comfortable with that estimate."

17   **compounds that have the unique capacity to keep muscle and vascular tissue pliant:** The role of the omega-3 fatty acid in salmon physiology is drawn from a 2009 interview with Frederic T. Barrows, research physiologist, USDA, Hagerman Fish Culture Experiment Station, Hagerman, ID.

18   **in advance of their spawning runs:** Overviews of salmon life cycles, migrations, and evolutionary history can be found in David R. Montgomery, *The King of Fish: The Thousand-Year Run of Salmon* (Cambridge, MA: Westview Books, 2003), and James A. Lichatowich, *Salmon Without Rivers: A History of the Pacific Salmon Crisis* (Washington, DC: Island Press, 1999).

18   **"The King of Fish":** "The mighty Luce or *Pike* is taken to be the *tyrant*, as the *salmon* is king, of the fresh waters," wrote Izaak Walton. Though his metaphorical intentions are open to interpretation, tryants and kings would have most certainly been on Walton's mind following his exile from London and Cromwell's execution of Charles II. Izaak Walton, *The Compleat Angler* (1653; Ithaca, NY: Cornell University Library, 2009).

18   **But in 1798:** Smaller dams were put across the Connecticut's tributaries before the dam at Turners Falls that undoubtedly reduced salmon populations. Turners Falls was, however, the first mainstem blocking. Salmon spawned below the Turners Falls dam, but the Turners Falls impediment was clearly the death knell to the run. For more background on Atlantic salmon on the Connecticut, see Montgomery, *The King of Fish*.

19   **after a handful of Danish and Faroe Islands fishermen:** A detailed account of Greenland catches made by Danish and Faroese fishermen is given in Anthony Netboy, *The Salmon: Their Fight for Survival* (Boston: Houghton Mifflin, 1974).

20   **California closed its salmon fishery:** The closure of the continental U.S. salmon fisheries was reported in "Salmon Fishing Closed for California, Oregon," *San Francisco Chronicle,* Aug. 11, 2008.

20   **human-controlled reproduction of Atlantic salmon:** A thorough summary of trends in European medieval fisheries, including a mention of the first salmon culture around the year 1400, is Richard C. Hoffman, "A Brief History of Aquatic Resource Use in Medieval Europe," *Helgoland Marine Research,* vol. 59, no. 1 (Apr. 2005), http://www.springerlink.com/content/69w8p244fu6lmwa2. It is also worth noting that much of early salmon culture, including significant hatchery operations in the United States and Europe in the nineteenth century, was designed to supplement declining or extirpated wild populations of salmon and trout, not to form the basis of commercial farms. Stock supplementation of many fish, particularly trout, is a common though controversial practice, and today most seemingly "wild" trout in the United States and Europe begin their lives in hatcheries.

20   **companies operating in the frigid fjords of southern Chile:** Salmon farming and wild salmon catch data, specifically tonnage and market share, are derived primarily from the World Wildlife Fund, *The Great Salmon Run: Competition Between Wild and Farmed Salmon,* ed. Gunnar Knapp, Cathy A. Roheim, and James L. Anderson, http://www.uri.edu/cels/enre/ENRE_Salmon_Report.html, 2007.

25   **Yukon king-salmon returns dropped far below:** The Alaska State Department of Fish and Game keeps careful records of salmon catches on a region-by-region basis and publishes them at http://www.cf.adfg.state.ak.us. Additional information on Alaska's salmon management methodology came from in-person interviews with Alaska Fish and Game departments in Emmonak, Alaska, as well as in the Bristol Bay salmon fishery. Howard Klein, then chairman of Ocean Bounty provided background on large-scale commercial use of salmon via an in-person interview in the summer of 2007.

26   **Before the Industrial Revolution, the world's population:** As the reader will see in my discussion of cod in chapter 3, it is extremely difficult to try to reconstruct historical fish populations, largely because habitat destruction and overfishing tend to have occurred before anyone has the impetus or means to count fish in the first place. Steve Gephard (cited above) wrote that while it is possible to look at catch histories of salmon, "(1) landings never equate to population size and (2) by the 1960s [when landing records start to become available] many salmon runs (like the Connecticut) had already been extirpated and other runs were decimated by dams, pollution, and overharvest in home waters. The North Atlantic Salmon Conservation Organization is working on a river database but it is not ready yet. It will include a list of all Atlantic salmon rivers in the world with some estimate of the amount of historical habitat. Once you have a list of all habitat, you could apply a theoretical production rate (e.g., 4 smolts per one production unit of habitat) and then a theoretical marine return rate (e.g., 0.01) and come up with a very rough estimate of the number of salmon historically. But we are years away from that." The prehistory of salmon is therefore often expressed as an inventory of different factors that caused salmon decline and an extrapolation of the probable loss of total fish. One such meta-analysis is in Frank Jensen, "Synopsis on the abundance of Atlantic salmon (Salmo salar L.) since the last ice age," Millenium Report of the Museum of Natural History (Aarhus, Denmark: Museum of Natural History, March 20, 1991). Jensen offers no numbers for the total historical population of salmon

but approximates a 99 percent decline from the last ice age through the present era with the steepest declines taking place in the wake of the Industrial Revolution.

33 **Jac Gadwill's socks:** Jac Gadwill left Kwik'pak not long after my June 2007 trip to Emmonak and no longer works for the company.

35 **"Thou shalt not let thy cattle breed with unlike animals":** Leviticus 19:19.

36 **"selective breeding":** My summary of animal breeding methodology is drawn primarily from Jay L. Lush, *Animal Breeding Plans* (Ames, IA: Iowa State University Press, 1937), and also from Trygve Gjedrem's *Selection and Breeding Programs in Aquaculture* (New York: Springer, 2005).

43 **piles of bright orange salmon fillets:** For an amusing and thorough account of the dot-com-like salmon-farming boom that started in Norway and spread to Chile, Canada, and elsewhere, see Aslak Berge, *Salmon Fever: A History of Pan Fish* (Bergen, Norway: Octavian, 2005).

44 **as salmon continue to be bred:** For discussions on feed-conversion ratios in farmed salmon and other aquaculture fish, see Rosamond L. Naylor, et al., "Feeding Aquaculture in an Era of Finite Resources," *Proceedings of the National Academy of Sciences of the United States of America*, vol. 106, no. 36 (Sept. 8, 2009), pp. 15,103–10, and Albert Tacon and Marc Metian, "Global Overview on the Use of Fish Meal and Fish Oil in Industrially Compounded Aquafeeds: Trends and Future Prospects," *Aquaculture*, vol. 285, no. 1–4 (2008), pp. 146–58. Published feed-conversion ratios can be misleading, due to the fact that farmers and scientists often use different measures for determining the weight of feed. Farmers tend to use dry weight, i.e., the weight of dried feed pellets fed to salmon, while ecologists tend to look at wet weight, i.e., the actual weight of the raw fish that went into making feed in the first place. A kilogram of dry feed is a dehydrated distillation of wet fish and represents a much larger amount of actual fish.

44 **displacing a self-sustaining wild fish population:** Numerous authors debate the genetic and pollution impact of salmon farms on wild salmon. A summary of the arguments against salmon farming can be found in the anthology *A Stain Upon the Sea: West Coast Salmon Farming*, by Stephen Hume, et al. (Madeira Park, BC, Canada: Harbour Publishing, 2004). For every argument against salmon farming, there is a phalanx of aquaculture scientists ready to dispute critics' claims. Both aquaculturists' claims and environmental concerns are presented in detail in Katherine Bostick, Jason W. Clay, and Aaron A. McNevin, *Aquaculture and the Environment: A WWF Handbook on Production Practices, Impacts, and Markets* (Washington, DC: Center for Conservation Innovation, World Wildlife Fund, 2005).

My impression from having looked at both sides of the debate is that diseases like infectious salmon anemia and parasites like sea lice represent the most palpable threat to salmon populations and that the genetic dilution of stocks is harder to prove. What is undeniable is that wild populations of Atlantic salmon are severely depressed and the severely diminished populations that remain are more vulnerable to disturbances in their environments than they would be if wild populations were abundant and robust.

49 **Diseases like infectious salmon anemia:** Infectious salmon anemia, or ISA, first appeared in the early 1990s and has risen and fallen in increasingly larger waves ever since. In 2010 Chilean salmon production dropped by a third as a result of ISA. Eduardo Thomson, "Chile Salmon Output to Fall a Third, Association Says," Bloomberg, Jan. 28, 2010.

52 **higher levels of PCBs:** The Pew-funded farmed-salmon-and-PCB study is: Ronald A. Hites, Jeffery A. Foran, David O. Carpenter, M. Coreen Hamilton, Barbara A. Knuth,

and Steven J. Schwager, "Global Assessment of Organic Contaminants in Farmed Salmon," *Science,* vol. 303, no. 5655 (Jan. 9, 2004), pp. 226–29.

55   **PCB contamination in farmed salmon may offset:** Mozaffarian's meta-analysis of the risks and benefits of eating fish is: Dariush Mozaffarian and Eric B. Rimm, "Fish Intake, Contaminants, and Human Health: Evaluating the Risks and the Benefits," *Journal of the American Medical Association,* vol. 296, no. 15 (Oct. 15, 2006), pp. 1885–99. In addition, an excellent summary of the debates around pollutants in fish and relative health benefits can be found in Marion Nestle, *What to Eat* (San Francisco: North Point Press, 2007).

67   **a modeling exercise conducted in 1998 by a consulting firm:** The study was conducted by ADI Ltd., 1998 and is described in *A WWF Handbook on Production Practices, Impacts, and Markets* as cited above.

69   **The world's very first aquaculturists:** A discussion of early aquaculture practices and how they could apply to a more environmentally benign approach can be found in: Barry Costa-Pierce, *Ecological Aquaculture: The Evolution of the Blue Revolution* (Oxford, UK: Wiley-Blackwell, 2002).

76   **The Donaldson is therefore a kind of genetic message in a bottle:** My source for salmon reintroduction on the Salmon River in New York State is Fran Verdoliva, Salmon River special assistant, New York State Department of Environmental Conservation. Verdoliva wrote me in the fall of 2009 that for the first time since the 1800s forty-seven naturally reproduced Atlantic salmon were found in tributaries leading into Lake Ontario. This is a most encouraging sign, since Atlantic landlocked salmon are the truly endemic salmon to Lake Ontario.

SEA BASS

81   **aquaculture is the fastest-growing food production system:** The UN Food and Agriculture Organization's report *The State of World Fisheries and Aquaculture 2008,* ed. J.-F. Pulvenis de Séligny, A. Gumy, and R. Grainger (Rome: FAO, 2009), gives the most recent statistics on the growth of aquaculture worldwide.

82   **striped bass—perhaps the most famous game fish:** An excellent account of the near demise and miraculous recovery of the American striped bass is Dick Russell, *Striper Wars: An American Fish Story* (Washington, DC: Island Press, 2005).

84   **The English word "bass" derives from:** Opinions differ on the derivation of bass. Anatoli Liberman had this to say, "I cannot say whether the origin of /barse/ was understood correctly by Friedrich Kluge, the author of a famous etymological dictionary of German, or James A. H. Murray, the great editor of the OED. Murray corresponded with German scholars on a regular basis. The first edition of Kluge and the first volume of the OED appeared almost simultaneously . . . but Kluge is a likelier candidate. Both refer barse/barsch to the root one has in English /bristle/, and both seem to have been right.

84   **Many moonfish are roundish and vaguely moonlike:** Any reader looking to play the fish name game could spend a useful hour exploring the University of British Columbia's Fish Base database, http://www.fishbase.org. Fish can be searched by common name or Latin name, and disambiguation information can further resolve conflicts on fish identity.

85   **Perciformes is the largest order of vertebrates on earth:** My summary of the perciform dilemma comes primarily from a 2008 interview with Joseph Nelson, author of the

frequently cited work on fish taxonomy *Fishes of the World,* 4th ed. (Hoboken, NJ: Wiley, 2006). One of the problems with sorting through the order Perciforme is that oceanic fish fossils typically fall to the bottom of the sea and then are reduced to magma when one continental plate is forced under another in the eons-long process of continental subduction. Contemporary taxonomists more and more are turning away from the fossil record and returning to fish classification armed with the tools of the relatively new discipline of phylogenetics. Phylogenetics compares living species' DNA and looks for evidence of common ancestry stored within DNA. A discussion of phylogenetic approaches to decoding and properly classifying the so-called basses and perciforms in general is Wm. Leo Smith and Matthew T. Craig "Casting the Percomorph Net Widely: The Importance of Broad Taxonomic Sampling in the Search for the Placement of Serranid and Percid Fish," *Copeia* 2007, No. 1.

86   **the fish we have come to recognize most widely as being edible:** My description of the relationship of the swim bladder to fish morphology derives primarily from a 2007 interview conducted with David L. G. Noakes, professor and senior scientist, Oregon Hatchery Research Center and Oregon State University. The evolutionary developments for swim bladders in fish are much older than the perciforms—dating back perhaps 250 million years, as opposed to the appearance of the perciforms 85 million years ago. Coastal perciforms, like European sea bass, have more limited pressure extremes and thus are more likely to be in a depth range reachable by a primitive fisher. It's of interest to note that benthic perciforms, like Chilean sea bass (cf. Patagonian toothfish), which live at depths exceeding two thousand feet, have smaller or even nonexistent swim bladders. Chilean sea bass/toothfish use oils secreted directly into tissues to meet their flotation needs. One reason that Chilean sea bass are so desirable as food fish is that their muscle tissues are infused with this flotation oil, making the fish extremely hard to overcook.

87   **the word *labros,* or "turbulence":** H. G. Liddell and Robert Scott, *An Intermediate Greek–English Lexicon* (Oxford, UK: Oxford University Press, 1945).

87   **"while with arts more exquisite the bass beguiles":** Several Roman poets writing on the European sea bass's cleverness are quoted in Jonathan Couch in *A History of the Fishes of the British Isles* (London: George Bell and Sons, 1848). Ovid notes the fish's tendency to burrow under a passing net, whereas Oppian suggests that the fish bends and twists and consciously makes a hole in its mouth to loose a hook. I don't discount that fish have specific behaviors when pursued or hooked, but I maintain that a fish's superior qualities are often anthropomorphic ascriptions and that the inability of fishermen to catch fish usually has more to do with a fish's abundance than with its skill at evading capture.

88   **as you go up the food chain each level is thinner:** My primary source for Mediterranean oligotrophia and interaction with human populations is a 2007 interview conducted with Constantinos C. Mylonas, Hellenic Centre for Marine Research, Iraklion, Greece.

89   **the meats they consumed consisted of:** Diets of Neolithic humans and Galton's principles for selection come from Juliet Clutton-Brock, *A Natural History of Domesticated Mammals* (New York: Cambridge University Press, 1999).

92   **nutrition to last the first few weeks:** Interviews with many aquaculture scientists contributed to my understanding of marine perciform domestication, including Constantinos Mylonas and Pascal Divanch at the Hellenic Centre for Marine Research in Iraklion, Greece; Josh Goldman at Australis Aquaculture, the cod farming research facilities at Fiskeriforskning and Akvaforsk in Norway; and Yonathan Zohar at the University of Baltimore.

94　**More than 70 percent of the fish Israelis ate were farmed:** A history of the early days of Israeli aquaculture can be found in "National Aquaculture Sector Overview: Israel," National Aquaculture Sector Overview Fact Sheets, text by J. Shapiro, FAO Fisheries and Aquaculture Department (online), Rome, updated July 6, 2006, http://www.fao .org/fishery/countrysector/naso_israel/en.

96　**marine aquaculture focusing on the Red Sea:** In addition to being directly quoted, Yonathan Zohar is also my primary source for the early development of aquaculture in Israel.

101　**Explosives tossed from the boats:** I have not encountered peer-reviewed documentation of Italians practicing dynamite fishing in the Ionian Sea, but Thanasis Frentzos maintains that the practice was a significant factor in the decline of wild sea bass in coastal Greece. Professor Konstantinos Stergiou, the director of the Laboratory of Ichthyology at the School of Biology of Aristotle University at Thessaloniki, confirmed the practice of dynamite fishing and the tendency of fishing methods to become more extreme the more reduced wild populations become.

103　**Mexican government would ban the United States from fishing for white sea bass:** A thorough description of issues surrounding white sea bass can be found in: Melissa M. Stevens, *White Sea Bass* (Monterey, CA: Monterey Bay Aquarium Seafood Watch, 2003).

103　**the Patagonian toothfish, that sold poorly:** Many authors (including the present one) have written about the Patagonian toothfish, aka Chilean sea bass. A book-length account of its discovery and naming is: G. Bruce Knecht, *Hooked: Pirates, Poaching, and the Perfect Fish* (Emmaus, PA: Rodale Books, 2007).

111　**digest themselves after death:** The use of rotifers and artemia as live feed is discussed at length in *Manual on the Production and Use of Live Food for Aquaculture*, FAO Fisheries Technical Paper 361, ed. Patrick Lavens and Patrick Sorgeloos, Laboratory of Aquaculture and Artemia Reference Center, University of Ghent, Ghent, Belgium, 1996.

112　**Von Braunhut invented a whole parallel universe:** Von Braunhut's unusual life and his marketing of "Sea Monkeys" is summarized in "Harold von Braunhut, Seller of Sea Monkeys, Dies at 77," *New York Times*, Dec. 21, 2003.

113　**"if we skimmed the oil off the top of the water":** Frentzos's innovation of skimming oil off the surface of rearing pens appears to have occurred at similar times at other research facilities throughout Europe.

118　**"Fish of Greece":** To my knowledge there is no thorough peer-review analysis of the genetic profiles of sea bass and sea bream in the Mediterranean before or after the introduction of farming. But the fact that farmed bass and bream now predominate over wild bass and bream in the Mediterranean is to most scientists, including Yonathan Zohar at the University of Maryland and Pascal Divanch at the Hellenic Centre for Marine Research at Iraklion, self-evident.

119　**the Rosetta stone of fish:** There is a case to be made that the red porgy (*Pagrus major*) in Japan was the Rosetta stone fish of ocean farming. Many of the developments with red porgy paralleled those of European sea bass. In fact, the conditions that motivated the Japanese to tame ocean fish were quite similar to those of Israel—an isolated nation deeply concerned about national food security and faced with diminishing marine resources. Nevertheless, it was European sea bass that scaled up the fastest and introduced an aquacultured ocean fish to a global market the soonest. It should also be noted that once sea bass culture was launched in the Mediterranean, a parallel program to domesticate gilthead sea bream (*Sparus aurata*) in the Mediterranean also took place, and many farms in Europe now cultivate sea bream and sea bass at the same time. Just as sea bass made their premiere in the American market under the Italian name branzino, sea bream have arrived in European clothes, often called by their Latin name "aurata" on contemporary menus. When aquaculturists speak of the great breakthroughs in marine fish culture that took place in the Mediterranean, they often speak of European sea bass and gilthead sea bream in tandem. At a certain point developments and breakthroughs with sea bass and sea bream occurred neck and neck.

COD

129   **"the last of wild food?":** My summaries of the history of codfish exploitation and the buildup of the codfish industry are drawn primarily from Mark Kurlansky, *Cod: A Biography of the Fish That Changed the World* (New York: Penguin, 1998) but also from a 2006 interview with George Rose, professor of fisheries at the Fisheries and Marine Institute of Memorial University Conservation at Memorial University, St. John's, Newfoundland, and a 2008 interview with Heike Lotze, chair in Marine Renewable Resources, Biology Department, Dalhousie University, Halifax, Nova Scotia.

132   **the price of cod had risen quickly:** The price of all wild fish fluctuates in the course of a season and over the course of years. Cod most certainly can be found on the market for $8 a pound, but an informal survey taken of fishmarkets in the New York City area in 2007 put the average price at around $13 a pound.

137   **And yet this assessment of stability is up for question:** The FAO lays out the issues of the reliability of its data mostly in reaction to a 2001 paper in the journal *Nature* by D. Pauly and R. Watson in "Fishery Statistics: Reliability and Policy Implications," FAO, 2002, http://www.fao.org/DOCREP/FIELD/006/Y3354M/Y3354M00.HTM.

137   **particularly in the United Kingdom:** British trends in seafood consumption comes from the Scientific Advisory Commission on Nutrition, "Advice on Fish Consumption, Benefits and Risks," Food Standards Agency and the Department of Health (Norwich, UK: Her Majesty's Stationery Office, 2004).

138   **Modern gadiforms evolved from the extinct genus *Sphenocephalus*:** My summary of gadiform evolution and radiation is derived primarily from Jurgen Kriwet and Thomas Hecht, "A Review of Early Gadiform Evolution and Diversification: First Record of a Rattail Fish Skull (Gadiformes, Macrouridae) from the Eocene of Antarctica, with Otoliths Preserved in Situ,"*Naturwissenschaften*, vol. 95, no. 10 (Oct. 2008), pp. 899–907, http://www.springerlink.com/content/b3262512uh182823.

142   **menu items like McDonald's Filet-O-Fish sandwich:** For those curious about in fast-food lore, it's interesting to note that the Filet-O-Fish sandwich was invented by Lou Groen, a McDonald's franchise owner in the Cincinnati area who found he was losing his largely Catholic customers on Friday because he had no fish item on the menu. The first Filet-O-Fish was made using halibut, causing the sandwich to cost around 30 cents (late 1960s prices). McDonald's executives demanded that Groen bring in the sandwich at 25 cents if they were to distribute it nationally. To bring in the sandwich at that price point, Groen turned to the much cheaper Atlantic cod. See Paul Clark, "No Fish Story: Sandwich Saved His McDonald's," repr. *USA Today*, Feb. 20, 2007.

145   **The United States had created a de facto marine reserve:** Approximately 17,000 square kilometers of Georges Bank, or 25 percent of the area, has been closed to bottom trawling. In addition to spurring a recovery of cod and other gadiforms, University of Rhode Island researchers noted a fourteen-fold increase in sea scallops. More information on Georges Bank recovery data and a map of closed areas can be found in Georges Bank Benthic Habitat Study, http://www.seagrant.gso.uri.edu/research/georges_bank/.

146 **To this day neither . . . has ever done such a thing:** This conclusion was made by Andy Rosenberg. Regulators in Europe and Canada would argue the semantics of this conclusion while acknowledging the inefficacy of regulation. Despina Pavlidou, director of the European Bureau for Conservation and Development within the Secretariat of the Intergroup on Climate Change and Biodiversity of the European Parliament, quite bluntly called the European Common Fisheries Policy "a failure."

146 **Georges Bank cod, the stock I was fishing:** The stock assessments and rebuilding targets for Georges Bank and Gulf of Maine cod derive primarily from interviews conducted with Loretta O'Brien and Ralph Mayo and their published paper: Loretta O'Brien and Ralph Mayo, *Status of Fishery Resources Off the Northeastern US: Atlantic Cod* (Woods Hole, MA: National Marine Fisheries Service Northeast Fisheries Science Center, Dec. 2006).

147 **the time horizon for rebuilding has been extended:** Rosenberg believes that the rebuilding target had to be extended for Georges Bank codfish because closure of fishing grounds did not occur fast enough back in the early 1990s. Had that last good spawning class of fish in the late 1980s not been fished so heavily, the biomass of the population might have been large enough to reach the earlier target date.

147 **the term "shifting baselines":** The original paper on the shifting-baselines theory is Daniel Pauly, "Anecdotes and the Shifting Baseline Syndrome of Fisheries," *Trends in Ecology and Evolution*, vol. 10, no. 10 (Oct. 1995), p. 430.

150 **status quo of scarcity:** Perhaps the most cited paper in the mainstream science press on the decline in fish abundance is Ransom Myers and Boris Worm, "Rapid Worldwide Depletion of Predatory Fish Communities," *Nature*, vol. 423, May 15, 2003, pp. 280–83.

152 **"cod have complex population structures":** Ted Ames's cod-population reconstructions can be found in: Edward P. Ames, "Atlantic Cod Stock Structure in the Gulf of Maine," *Fisheries*, vol. 29, no. 1 (Jan. 2004), pp. 10–28.

152 **"an awful lot of cod":** One anecdotal, personal observation on the recovery of cod: In the last two years, cod have appeared off Montauk, New York, in significant numbers for the first time in nearly two decades. In the winter of 2010, recreational fishing boats from western Long Island from as far west as New York City relocated to Montauk to get in on the fishery. Even boats in the Greater New York area reported catches of fifty to sixty codfish per boat in February of 2010. Biologists I interviewed were reluctant to say whether this newfound abundance in southerly waters represented a long-term trend.

154 **"lobster glut":** The recovery of lobsters in Maine is detailed in Melissa Clark, "Luxury on Sale: The Lobster Glut," *New York Times*, Dec. 10, 2008, p. D3. Nevertheless some marine biologists have argued that the boom in lobsters (and snow crabs) owes itself not to good management of lobsters but rather to a dearth of predators like codfish that once preyed on juvenile lobsters. Ames disputes this point, arguing that accounts from colonial New England record massive amounts of cod *and* lobster being present in coastal fisheries.

157 **fly the Norwegian flag nearly as often as they do the Union Jack:** Observations on the Shetland Islands economy and social structure are based on interviews I conducted in and around Lerwick in the early spring of 2007.

158 **laws that required cod be granted the "five freedoms":** The five freedoms are detailed at http://www.fawc.org.uk/freedoms.htm.

159 **This wild cod bacchanalia is an annual ritual:** The information about the interplay between wild *skrei* cod and farmed cod comes from interviews I conducted in the spring of 2006 with officials from the Norwegian Seafood Export Council, Tromso, Norway.

166   **Things at No Catch started going downhill in 2008:** The implosion of No Catch's farmed-cod attempt is detailed in Severin Carrell, "World's First Organic Cod Farm Sinks into Administration with £40m Debt," *Guardian*, Mar. 6, 2009.

169   **But Unilever managed to pull off:** An account of the relationship of Unilever, Greenpeace, the World Wildlife Fund, and the Marine Stewardship Council appears in: Bob Burton, *Inside Spin: The Dark Underbelly of the PR Industry* (Sydney: Allen & Unwin Academic, 2008).

176   **The fish that were in Mr. Khon's pond:** Most information about the Vietnamese *Pangasius* industry comes from a May 2008 research trip up the Mekong River. Particularly useful were the accounts of Flavio Corsin, an aquaculture scientist affiliated with the World Wildlife Fund's Aquaculture Dialogues, who has been a resident in Vietnam throughout the explosion in *Pangasius* culture. Statistics on growth of *Pangasius* production can be found at http://www.worldwildlife.org/what/globalmarkets/aquaculture/dialogues-pangasius.html.

180   **name for itself in the abundance arena:** Information on tilapia reproduction and growth is derived from e-mail interviews with Ron Phelps, assistant professor, Department of Fisheries and Allied Aquacultures, University of Alabama at Auburn.

181   **opportunity to turn the fish into a moneymaker:** Information about tilapia culture derives primarily from a session of the World Wildlife Fund's Tilapia Aquaculture Dialogues I attended in December 2008.

181   **off-flavor is one of the key reasons:** Information on off-flavor came to me primarily through interviews with members of the catfish industry in Mississippi, Arkansas, and Alabama in the summer of 2008, with a detailed contribution coming from Craig Tucker, director, National Warmwater Aquaculture Center, Mississippi State University.

185   **the considerably less regulated coast of Russia:** Alaska pollock's drift into Russian waters was reported in Kenneth R. Weiss, "U.S. Fishing Fleet Pursues Pollock in Troubled Waters," *Los Angeles Times*, Oct. 19, 2008.

TUNA

192   **Tuna in the western Atlantic follow the river of higher-temperature water:** Background on tuna morphology, migratory patterns, and hunting behavior comes primarily from Carl Safina, *Song for the Blue Ocean: Encounters Along the World's Coasts and Beneath the Seas* (New York: Henry Holt, 1998), and Richard Ellis, *Tuna: A Love Story*, (New York: Alfred A. Knopf, 2008).

194   **an article called "The Holy Tuna Tablets" maintains:** "The Holy Tuna Tablets" can be accessed at http://www.screamingreel.com/HolyTunaTablets.

199   **Their range encompasses nearly the entirety of the ocean:** Transoceanic migratory patterns of bluefin are still very much under research. In the Pacific, bluefin spawn on the western side of the ocean with a portion of juveniles migrating to the east. In the Atlantic, bluefin spawn in both the Gulf of Mexico and the Mediterranean. Though the eastern Atlantic and western Atlantic bluefin are considered two different stocks, there is documented mixing of the population on both sides of the Atlantic and it is speculated that the mixing of the western and eastern stocks is an important factor in maintaining the overall health of the Atlantic bluefin population. For Pacific migratory patterns see Jeffrey J. Polovina, "Decadal variation in the trans-Pacific migration of northern bluefin tuna (Thunnus thynnus) coherent with climate-induced change in prey abundance," *Fisheries Oceanography*, vol. 5, no. 2 (Oct. 5, 2007). For Atlantic migratory patterns see Barbara A. Block et al., "Electronic Tagging and Population Structure of Atlantic Bluefin Tuna," *Nature*, vol. 434128 (April 2005).

201   **Catches from the high seas have doubled:** Data on high-seas catch trends were obtained from Wilf Swartz, a researcher at the University of British Columbia Fisheries Centre.

201   **tuna sushi is a relatively new invention:** Trevor Corson, the author of the highly entertaining and informative *The Story of Sushi: An Unlikely Saga of Raw Fish and Rice* (New York: Harper Perennial, 2008), provided information on the history of tuna and sushi in Japan, as well as some of the biochemical explanations as to why sushi is appealing to those who usually eschew cooked fish. In answering my questions about Japanese sushi habits, Corson drew on several sources from the Japanese translated by Corson and Sakiko Kajino. They are: *Nihonjin wa sushi no koto o nani mo shiranai* [The Japanese Know Nothing About Sushi], ed. Mitsuru Nakamura (Tokyo: Gakken, 2003); Morihiko Sakaguchi, Michiyo Murata, Satoshi Mochitzuki, and Yoshihiro Yokoyama, *Sakana hakase ga oshieru sakana no oishisa no* [Fish Experts Teach the Secrets of the Deliciousness of Fish] (Tokyo: Hamano Shuppan, 1999); Shinzo Satomi, *Sukiyabashi Jirō shun o nigiru* [Jirō of Sukiyabashi Makes Sushi with the Seasons] (Tokyo: Bungei Shunjū, 1997); and Masuo Yoshino, *Sushi, Sushi, Sushi: Sushi no Jiten* [Sushi, Sushi, Sushi: The Encyclopedia of Sushi] (Tokyo: Asahiya Shuppan, 1990).

202   **sportfishing of giant, thousand-pound Atlantic bluefin tuna:** A full account of the bluefin fishery in Canada and the Japanese businessmen who brought Canadian bluefin to Japan can be found in Sasha Issenberg, *The Sushi Economy: Globalization and the making of a Modern Delicacy* (New York: Gotham, 2007).

205   **"... I'd catch all of them if I could":** Steven Weiner, a bluefin harpooner, is quoted in John Seabrook, "Death of a Giant," *Harper's*, June 1994.

212   **"... orderly development of the whaling industry":** My summaries of the history of whale taxonomy and portions of the history of the whale-conservation movement were drawn primarily from interviews with D. Graham Burnett and his book *Trying Leviathan: The Nineteenth-Century New York Court Case That Put the Whale on Trial and Challenged the Order of Nature* (Princeton, NJ: Princeton University Press, 2007).

212   **"... directly related to problems that I, as a biologist":** Quotes from Roger S. Payne came from: Amy Standen, "Roger S. Payne," *Salon*, Oct. 30, 2001, http://www.salon.com/people/bc/2001/10/30/roger_payne.

216   **research subjects end up in restaurants as whale carpaccio:** Background on the Norwegian whaling trade was obtained through interviews with Phillip Clapham, research fisheries biologist and vice president, Center for Cetacean Research & Conservation.

217   **total number of giant bluefin spawners:** My reference for the current status of the western Atlantic bluefin tuna stock is Carl Safina and Dane H. Klinger, "Collapse of Bluefin Tuna in the Western Atlantic," *Conservation Biology*, vol. 22, no. 2 (April 2008), pp. 243–46. Fishermen, of course, dispute Safina's and others' grim bluefin assessments, but Safina is quick to point out that fishermen in the United States were only able to catch 10 percent of their allowable quota in 2006. Either bluefin were too smart to get caught (highly unlikely), or there simply weren't enough of them to justify the allowable catch.

217   **"... further reduction in spawning stock biomass":** ICCAT's assessment of bluefin tuna stocks can be found in "Stock Status Report 2008: Northern Bluefin Tuna—East Atlantic and Mediterranean Sea," http://firms.fao.org/firms/resource/10014/en.

220   **for the first time both the United States and the European Union backed a CITES listing for bluefin:** In addition to bluefin, conservationists also pushed for four species of threatened shark to be included in CITES Appendix II. Sharks are increasingly victim to "finning," in which they are caught, stripped of their fins (for shark-fin soup),

and dumped overboard. In the end three of the four sharks failed even to make it past the committee phase of the CITES process. One shark, the probeagle, was approved in committee by a single vote margin but then rejected by the end of the final plenary session. It is not just tuna that have a hard time being wildlife. It seems as if all fish suffer from the same discrimination.

222 **a remarkable recovery for swordfish:** A full account of the Atlantic swordfish rebuilding process can be found at John Pickrell, "North Atlantic Swordfish on Track to Strong Recovery," National Geographic News, Nov. 1, 2002, http://news.national-geographic.com/news/2002/11/1101_021101_Swordfish.html.

224 **Minimata disease:** An account of the Minamata mercury poisoning can be found in: Masazumi Harada, "Environmental Contamination and Human Rights—Case of Minamata Disease," *Organization & Environment*, vol. 8, no. 2 (1994), pp. 141–54.

225 **mercury concentrations amplify in fish at higher levels on the food chain:** As with PCBs, readers seeking a more detailed explanation of mercury contamination and contamination-level standards would once again benefit by referring to Marion Nestle's excellent summary in *What to Eat* (San Francisco: North Point Press, 2007).

225 **"Consumers should not be misled that a system of management":** A detailed critique of seafood-choice campaigns and their effects on policy is: Jennifer L. Jacquet and Daniel Pauly, "The Rise of Seafood Awareness Campaigns in an Era of Collapsing Fisheries," *Marine Policy*, vol. 31 (2007), pp. 308–13.

227 **the first large-scale captive spawning of tuna:** An exhaustive and fascinating description of the domestication of bluefin tuna can be found in Richard Ellis's *Tuna: A Love Story*.

234 **have decided to call the fish "Kona Kampachi":** Kahala / Kona Kampachi are one of several species of the family Seriolla under cultivation at the present time. Japan has a long tradition of growing yellowtail (hamachi in sushi parlance), but the industry is still based upon capture of animals from the wild. Australians are also growing yellowtail in large numbers. What stood out for me with Sims's operation is his commitment to proper siting of his farms, his quest to find feeds that are low in fish oil and fish meal, and the use of a species that is in high abundance in the wild because there is no commercial value for its wild form.

234 **"Kona Kampachi, that's an artificial name":** The sushi chef quoted is the brilliant Naomichi Yasuda, the creator of Sushi Yasuda in Manhattan.

240 **But the final gear in the system, the tuna, the part that interested me most, was missing:** Those who follow the bluefin fishery closely will no doubt note that while bluefin fishing in 2009 was terrible, in the spring of 2010, shortly before this book went to press, an outstanding run of bluefin tuna occurred off North Carolina, the likes which had not been seen in many years. However, according to researchers with the Tag a Giant Foundation who have tagged and studied these fish for more than a decade, this burst of fish belongs to an unusually good year class of fish that is a spike in an otherwise downward population trend. In graphs of fisheries declines for many species, these sorts of spikes are common, occurring on a sometimes decade-long interval and attributable to a good year of spawing conditions or juvenile survival. But in a classic fisheries decline the population peaks get lower and lower, as do the valleys, and many a fisheries regulator has been duped by these false peaks.

Nearly all the fish off Cape Hatteras in the spring of 2010 were "small" fish of two hundred pounds or less, just under the commercial-size limit. If this good year class is protected, they could indeed form the basis of a larger, species-wide recovery. But in just a few months these fish will be big enough for commerical boats to target, and should they be wiped out, successive peaks and valleys will go lower still.

241   **"in the early days of the founding of the United States":** These words were spoken by Joseph Powers, former head of the scientific committee of ICCAT, and refer to the "three-fifths compromise" struck between Northern and Southern states whereby slaves would be counted at three-fifths of their numbers for representation purposes.

CONCLUSION

244   **the Monterey Bay Aquarium . . . took the brave act of commissioning a survey of the program's effects:** The survey referred to is Quadra Planning Consultants Ltd. (2004) Seafood Watch Evaluation: Summary Report, Galiano Institute for the Environment, Salt Spring Island.

245   **"gifts of the sea":** The Russian term for seafood is дары моря (*dary morya*), "gifts of the sea," although the more technical Soviet term морепродукты (*moryeprodukty*), "sea products," may also be used.

246   **ocean acidification is a real and growing threat:** According to a January 16, 2009, article in *Science*, fish have the ability to produce calcium carbonate, a substance that makes seawater pH more basic. As much as 15 percent of the ocean's calcium carbonate may originate from fish wastes. See R. W. Wilson, "Contribution of Fish to the Marine Inorganic Carbon Cycle," *Science*, Jan. 16, 2009.

246   **The world fishing fleet is estimated by the United Nations:** Statistics on fisheries subsidies can be found in Rolf Willmann and Kieran Kelleher, eds., *The Sunken Billions: The Economic Justification for Fisheries Reform* (Washington and Rome: World Bank and UN Food and Agriculture Organization, 2008).

247   **no-catch areas:** The concept of Marine Protected Areas is nearly universally loved by conservationists and reviled by fishermen. The "spillover" effect of no-catch areas is hotly debated, and many fishermen contend that there is not enough science to justify the closure of fishing grounds. I lay out my argument for marine protected areas in more detailed form in Paul Greenberg, "Ocean Blues," *New York Times Magazine*, May 13, 2007. More recent data is just coming in. In February of 2010 a panel of scientists tracking the Great Barrier Reef's no-catch areas over the last five years concluded that fish populations were usually double the level of those in nonfished areas of the reef.

253   **fish should not be farmed too densely:** In a fall 2009 interview, Dr. Paddy Gargan, senior research scientist with the Central Fisheries Board of Ireland, stressed repeatedly the negative impacts of sea lice on wild salmon populations. At the same time, Gargan noted that salmon farms sited in open sea areas away from the migration lanes could potentially have much less damaging effects on wild populations.

INDEX

ABOUT THE AUTHOR

Paul Greenberg is the author of *Leaving Katya: A Novel* and a frequent contributor to the *New York Times Magazine*, the *New York Times Book Review*, and the *Times'* Opinion page. A National Endowment for the Arts Literature Fellow as well as a W. K. Kellogg Foundation Food and Society Policy Fellow, he lives and works in New York City and Lake Placid, New York.